Joel Ochoa

CRYPTOCURRENCY

The Digital Monetary Revolution

Joel Ochoa

Cryptocurrency the digital monetary revolution

Joel Ochoa

DEDICATION

Dedicated to all pioneers in new technologies existing and to come

Cryptocurrency the digital monetary revolution

Joel Ochoa

CONTENTS

Cryptocurrency the digital monetary revolution

Joel Ochoa

THANKS

I am grateful for this time in history and allowing me to publish this book through the internet.   I would not have believed it in 1990 if someone had told me   just to imagine the new technologies to come makes me smile.

Joel Ochoa

1 THE ORIGINS OF CRYPTOCURRENCIES

In this chapter, we will begin the story behind the invention of cryptocurrencies, starting with the creation of Bitcoin by Satoshi Nakamoto in 2008. We will discuss how this innovative blockchain technology laid the foundation for the emergence of a new financial system.

Bitcoin was the first successful implementation of a cryptocurrency based on blockchain technology. This revolutionary technology has allowed the transfer of value in a completely decentralized and transparent way, without relying on intermediaries such as traditional banks.

The blockchain is a public and distributed registry in which all transactions made with a particular cryptocurrency, such as Bitcoin, are recorded. Explore how this system works, which is based on the validation of transactions by a network of nodes or miners.

In Bitcoin's underlying technology, blocks play a critical role in its operation. Each block is a unit of information that contains a set of Bitcoin transactions. These transactions are recorded chronologically and secured using cryptographic techniques.

Each time a Bitcoin transaction is made, it is bundled with other pending transactions in a block. Miners, who are network participants who use their processing power to solve complex mathematical problems, compete to find a solution to the problem. The first miner to find the solution validates and adds the block to the existing blockchain. In exchange for their work, the miner receives a reward in the form of new bitcoins, as well as transaction fees associated with transactions included in the block.

The blockchain, or blockchain, is the structure that is created as blocks are added sequentially. Each block contains a unique identifier, known as a hash, which is generated from the data of the previous block, the transactions included in the block, and other metadata. This ensures that any changes to a previous block would affect all subsequent blocks, making the chain immutable and resistant to tampering.

The Bitcoin blockchain is public and accessible to all network participants. Each node in the network has a copy of the complete blockchain, ensuring that all participants have access to the same information. This provides transparency and security, as Any attempt to alter a transaction would require a great deal of computational power and the consensus of most network participants.

In short, blocks in Bitcoin contain a set of transactions that are aggregated chronologically and secured using cryptography. These blocks form the Bitcoin blockchain, which securely records all transactions made with cryptocurrency. The blockchain is immutable and tamper-resistant, providing transparency and security in the Bitcoin system.

We will talk about the key concepts of cryptography, which is used to ensure the privacy and integrity of transactions on the blockchain. We will discuss how digital signatures and cryptographic keys are used to ensure that only legitimate owners can access and transfer their cryptocurrencies.

In addition, we will discuss the role of miners in the blockchain network. We will explain how miners use their computational power to solve complex mathematical problems and add new blocks of transactions to the chain. We will discuss the concept of consensus, which is fundamental to ensuring the integrity of the blockchain and preventing fraud.

To better illustrate these concepts, here's a simplified example: Imagine you have a blockchain on which all Bitcoin transactions are recorded. Each block contains a set of transactions and a reference to the previous block. Each time a new transaction is made, miners compete to solve a mathematical problem related to the current block. The first miner to solve the problem adds a new block to the chain and receives a reward in Bitcoin for his work. Once the block is added, transactions are considered confirmed and cannot be modified, ensuring the security and integrity of the block.

In summary, in this section we will explore how the birth of Bitcoin and blockchain technology laid the foundation for the emergence of cryptocurrencies. We will discuss the decentralization, transparency, and security that blockchain provides, as well as the essential role of cryptography and miners in the operation of this revolutionary technology.

After the creation of Bitcoin all other cryptocurrencies are usually called Altcoins, here is a list of the first ten created after Bitcoin currently in circulation in June 2023 there are approximately in circulation: 25,404 altcoins including tokens

Here is a list of the first 10 cryptocurrencies, altcoins or tokens created after Bitcoin, along with a brief explanation of each of them:

1. Litecoin (LTC): Created in 2011, Litecoin was one of the first cryptocurrencies to emerge after Bitcoin. It was developed with the aim of improving the speed of transactions and offering a more efficient alternative compared to Bitcoin.

2. Namecoin (NMC): Launched in 2011, Namecoin was the first cryptocurrency that sought to provide a decentralized and censorship-resistant domain name system. Its goal was to allow the creation of web domains without relying on centralized intermediaries.

3. Peercoin (PPC): Introduced in 2012, Peercoin is based on a combination of proof-of-work and proof-of-stake, making it one of the first cryptocurrencies to implement this mechanism. Their goal was to achieve greater energy efficiency and security compared to Bitcoin.

4. Ripple (XRP): Created in 2012, Ripple focused on facilitating fast and inexpensive international payments and money transfers. Unlike most cryptocurrencies, Ripple does not use a public blockchain, but instead employs its own consensus protocol to validate transactions.

5. Dogecoin (DOGE): Born in 2013 as a joke cryptocurrency, or a "MEME COIN," Dogecoin quickly gained popularity on social media and became a "meme coins" phenomenon. Although initially lacking a specific use, it has been used to make donations and as a form of online gratitude.

6. Nxt (NXT): Launched in 2013, Nxt was one of the first cryptocurrencies to implement a proof-of-stake (PoS) system instead of the proof-of-work (PoW) used by Bitcoin. In addition to being a cryptocurrency, Nxt also offered a platform for the development of decentralized applications (dApps).

7. Mastercoin (OMNI): Created in 2013, Mastercoin is considered one of the first cryptocurrencies that introduced the concept of tokens on the blockchain. It allows the creation and management of custom tokens, which has paved the way for the development of numerous token-based projects.

8. Counterparty (XCP): Based on the Bitcoin blockchain, Counterparty was launched in 2014 as a platform that allowed the issuance and exchange of tokens. It also offered features like smart contracts and decentralized betting options.

9. BitShares (BTS): Created in 2014, BitShares is a blockchain platform that focuses on decentralized financial services, such as decentralized exchanges, asset issuance, and smart contracts. It seeks to offer a solid infrastructure for online financial applications.

10. Ethereum (ETH): Launched in 2015, Ethereum is one of the most important and well-known cryptocurrencies after Bitcoin. It stands out for allowing the execution of smart contracts and the development of decentralized applications.

. Ethereum has been instrumental in the rise of ICOs (Initial Coin Offerings) and has driven the adoption of smart contracts in the cryptocurrency world.

Importantly, some of these cryptocurrencies have had a significant impact and proven useful in various use cases, while others have gained popularity as "meme coins" without a specific purpose or use.
We must also mention mining and its importance in both Bitcoin and other Altcoins.
Bitcoin mining is the process by which new bitcoins are created and all transactions are recorded on the Bitcoin network. It is an essential part of the functioning of cryptocurrency and allows us to maintain the integrity and Network Security.

Next, I explain the key concepts of Bitcoin mining:

1. **Blocks and Blockchain: ** Bitcoin transactions are grouped into blocks. Each block contains a set of confirmed transactions that have been made on the network. The blocks are linked together in chronological order to form a continuous chain of blocks, known as the blockchain.

2. **Hashing and Proof of Work (PoW):** Each block in the blockchain has a unique code called a "hash" that is generated using a special cryptographic function. To add a new block to the chain, miners must compete to solve a complex mathematical problem known as "proof of work" (PoW). Miners try to find a hash that meets certain predefined requirements. This PoW process requires a lot of computing power and is commonly referred to as "mining".

3. **Miner Reward: ** The miner who solves the PoW problem and successfully adds a new block to the blockchain receives a reward in newly created bitcoins, as well as transaction fees paid by users to include their transactions in the block. This reward in bitcoins is how new bitcoins are generated and introduced into circulation.

4. **Mining Difficulty: ** The difficulty of the PoW problem is automatically adjusted to maintain a constant rhythm of block creation on the network. When there is more computing power in the network (more miners), the difficulty Increase so that the blocks occur approximately every 10 minutes on average.

5. **Miner Competition: ** Bitcoin mining is a competition between miners, as only one of them can Add a new block to

the blockchain and receive the reward. This creates a decentralized system in which no individual or entity has control over the generation of new bitcoins or the validation of transactions.

Bitcoin mining is critical to ensuring the security and reliability of the network, as well as creating new bitcoins and processing transactions. However, the mining process requires a great deal of energy and computing power, which has led to debates about its sustainability and its impact on the environment.

Let's See Also Bitcoin Halving

Bitcoin halving is a pre-programmed and periodic event that occurs approximately every four years on the Bitcoin network. During halving, the reward miners receive for validating and confirming blocks of transactions is halved.

In the Bitcoin system, miners are responsible for aggregating and verifying transactions in blocks to add them to the Bitcoin blockchain. In exchange for their work and the use of their computing power, miners receive a reward in the form of new bitcoins. This reward is a way to "create" new bitcoins and is called a "block reward".

The Bitcoin protocol states that the block reward will be halved every 210,000 blocks mined, which happens roughly every four years. At the time of Bitcoin's launch in 2009, the block reward was 50 bitcoins. After the first halving in 2012, The reward was reduced to 25 bitcoins, and after the second halving in 2016, it was reduced to 12.5 bitcoins.

Joel Ochoa

The last Bitcoin halving occurred in May 2020, where the block reward was reduced from 12.5 bitcoins to 6.25 bitcoins. The next halving is expected to occur around the year 2024.

Halving has a significant impact on Bitcoin's economy and its circulating supply. By reducing the block reward, the rate of creation of new bitcoins decreases, resulting in a decrease in the inflation rate of Bitcoin. Over time, this leads to a relative shortage of new bitcoins in circulation and can have a bullish effect on the price of Bitcoin.

The halving is considered an important event for the Bitcoin community and for investors, as it can have a significant impact on the supply and demand of bitcoins, which can affect the price of the cryptocurrency in the short and long term. However, it is also important to note that the price of Bitcoin is influenced by a variety of factors, including market demand, adoption, regulations, and the general perception of its value as a digital asset.

To roughly project when the last Bitcoin mined block will occur and when all coins are in circulation, we can use the available information about the rate of creation of new bitcoins through halving's.

Bitcoin's third halving occurred in May 2020, reducing the block reward from 12.5 bitcoins to 6.25 bitcoins. The total number of bitcoins that can be created is 21 million, and it is estimated that all bitcoins will be in circulation approximately in the year 2140.

The rate of issuance of new bitcoins decreases by half with each halving, which means that every four years, half of the number of bitcoins that were created in the previous period is created. Here is a rough projection based on the history of halving's:

- The first halving occurred in November 2012, reducing the block reward from 50 bitcoins to 25 bitcoins.
- The second halving occurred in July 2016, reducing the block reward from 25 bitcoins to 12.5 bitcoins.
- The third halving occurred in May 2020, reducing the block reward from 12.5 bitcoins to 6.25 bitcoins.

Based on these periodic block reward reductions and considering that the rate of creation of new bitcoins decreases by half approximately every four years, the last mined block of Bitcoin and all coins is expected to be in circulation around the year 2140. It is important to note that these estimates may vary due to external factors and changes in the Bitcoin mining rate, but to date, this projection remains a reasonable approximation.

The Satoshi

The "satoshi" are the smallest unit of measurement of Bitcoin, the most well-known and used cryptocurrency in the world. It was named after its creator, Satoshi Nakamoto, who is the pseudonym under which The Bitcoin whitepaper and the network was launched in 2009.

Bitcoin is divided into smaller units to facilitate transactions of lower value and allow for greater accuracy in payments. One Bitcoin (BTC) can be divided into 100 million satoshis, which means that one satoshi represents one hundredth millionth of a Bitcoin. The unit of measure was denominated in honor of its creator to honor his contribution to the creation of the Bitcoin protocol.

Therefore, if you have 1 Bitcoin, you will have a total of 100 million satoshis. Similarly, if you have 0.001 Bitcoin, you will have 100,000 satoshis.

The use of satoshis as a unit of measurement is particularly relevant in the context of microtransactions or small payments, as it allows for more accurate and efficient split transactions.

In short, satoshis are the smallest units into which a Bitcoin can be divided, and their name pays homage to Satoshi Nakamoto, the mysterious creator of Bitcoin and the blockchain protocol on which this cryptocurrency is based.

What other coins can be mined?
Apart from Bitcoin, there are many other cryptocurrencies that can be mined using different consensus algorithms. Some of the most popular cryptocurrencies that still support mining are as follows:

1. Ethereum (ETH): Ethereum is the second largest cryptocurrency by market cap and is based on a consensus algorithm called Proof of Work (PoW). However, Ethereum is in the process of switching to a Proof-of-Stake (PoS) consensus system called Ethereum 2.0, which will eventually eliminate PoW mining.

2. Litecoin (LTC): Litecoin is a cryptocurrency that was created as an alternative to Bitcoin and uses the Script consensus algorithm for mining.

3. Bitcoin Cash (BCH): Bitcoin Cash is a fork of Bitcoin that was created to address Bitcoin's scalability issues. It uses the SHA-256 consensus algorithm, similar to Bitcoin, for mining.

4. Monero (XMR): Monero is a privacy-focused cryptocurrency that uses the CryptoNight consensus algorithm for mining.

5. Dash (DASH): Dash is a cryptocurrency with a focus on transaction speed and privacy. It uses the X11 consensus algorithm for mining.

6. Zcash (ZEC): Zcash is another privacy-focused cryptocurrency and uses the Equihash consensus algorithm for mining.

7. Dogecoin (DOGE): Dogecoin is a cryptocurrency that started as a joke, but has gained popularity over the years. It is based on the Scrypt algorithm and can be mined.

8. Ravencoin (RVN): Ravencoin is a cryptocurrency designed to facilitate the transfer of digital assets. It uses the X16R algorithm for mining.

These are just a few of the many cryptocurrencies that can be mined today. Importantly, the mining process may vary depending on the cryptocurrency and consensus algorithm used. Some cryptocurrencies, like Ethereum, they are in the process of changing their consensus system, which may affect how they are mined in the future.

Which ones cannot be mined?

Some cryptocurrencies cannot be mined because they use a different consensus mechanism than traditional mining, such as Proof of Stake (PoS) or Proof of Authority (PoA). These mechanisms do not require miners to compete to solve complex mathematical problems, as in Bitcoin mining. Instead, they use other forms of network validation and security. Below, some of the cryptocurrencies that cannot be mined are:

1. **Ethereum 2.0 (ETH):** Ethereum, the second largest cryptocurrency by market cap, is in the process of migrating from a Proof-of-Work (PoW)-based mining system to a Proof-of-Stake (PoS)-based one on Ethereum 2.0. When the migration is complete, Ethereum will no longer be minable, and participants will be able to validate and secure the network by staking their ETH.

2. **Cardano (ADA):** Cardano is a blockchain platform that uses the Ouroboros PoS consensus algorithm. ADA holders can participate in staking to validate transactions and secure the network, rather than mining blocks.

3. **Polkadot (DOT):** Polkadot is an interoperability network that uses the Nominated Proof of Stake (NPoS) consensus algorithm. DOT holders can participate in staking to contribute to network security and governance.

4. **Binance Coin (BNB):** Binance Coin, the native

cryptocurrency of the Binance exchange, is based on the Binance Chain Tendermint consensus algorithm. BNB holders can participate in staking to contribute to network security and earn rewards.

5. **Cardano (ADA):** Cardano is a blockchain platform that uses the Ouroboros PoS consensus algorithm. ADA holders can participate in staking to validate transactions and secure the network, rather than mining blocks.

6. **Tezos (XTZ):** Tezos uses the Liquid Proof of Stake (LPoS) consensus algorithm. XTZ holders can participate in staking to validate blocks and secure the network.

These are just some of the cryptocurrencies that cannot be mined and use other consensus mechanisms, such as staking, to ensure the security and functioning of the network. Each cryptocurrency can have its own consensus system and specific rules, allowing them to operate efficiently and securely.

2 WHAT ARE CRYPTOCURRENCIES FOR?

Of course! Let's explain cryptocurrencies and their current use. Cryptocurrencies are digital currencies that use cryptography to ensure secure transactions and control the creation of new units. As they have evolved, they have demonstrated various uses in different domains. Here's an overview. of some of the main use cases of cryptocurrencies today:

1. Medium of exchange: Cryptocurrencies, such as Bitcoin and Litecoin, are increasingly used as a medium of exchange to purchase goods and services online and, in some cases, also in physical establishments. They provide an alternative to traditional fiat currencies and allow fast and secure transactions without the need for intermediaries.

2. Investment and speculation: Many people have started investing in cryptocurrencies as a way to diversify their portfolio and look for potentially high returns. The volatility of cryptocurrencies has attracted investors and speculators looking to take advantage of price changes.

3. International transfers: Cryptocurrencies, such as Ripple and Stellar, offer efficient solutions for international money transfers. By cutting out middlemen and using distributed ledger technologies, cryptocurrencies enable fast and inexpensive transactions without the complications and high costs associated with traditional transfer methods.

4. Smart contracts: Ethereum, along with other platforms such as Cardano and EOS, has driven the development of smart contracts. These contracts are autonomous software that runs automatically when certain pre-established conditions are met. Smart contracts have applications in various fields, such as digital asset management, decentralized marketplaces, and financial applications.

5. Asset tokenization: Cryptocurrencies have enabled the tokenization of physical and digital assets. This involves the representation of an asset in the form of a token on the blockchain, making it easier to exchange and transfer. Tokens can represent anything from traditional assets such as real estate and artwork to access rights to services or even fractions of larger assets.

6. Decentralized Finance (DeFi): Cryptocurrencies have given rise to a boom in decentralized finance (DeFi), which are platforms and protocols that allow people to access and participate in a wide range of financial services without intermediaries. Users can make loans, earn interest, exchange assets, and participate in project governance, among other activities.

These are just a few examples of the current uses of cryptocurrencies. As technology and adoption continue to advance, new use cases and innovative applications are likely to emerge in different industries. Cryptocurrencies are revolutionizing the way we think about money and value transfer, bringing new opportunities and challenges in the financial world and beyond.

Other examples of use of Altcoins is Staking through which you can generate income in the form of interest as if you lent your coins and in doing so you make a profit.

Staking is a process whereby holders of certain cryptocurrencies lock their assets on a specific wallet or platform to support the operations of a red blockchain. In exchange for blocking their funds, staking participants receive rewards in the form of more cryptocurrency. This practice is based on the consensus mechanism called "proof of stake", which is an alternative to the traditional "proof of work" used by Bitcoin.

When participating in staking, users contribute to the security and efficiency of the blockchain network by keeping a certain amount of cryptocurrency in their wallet or on a staking platform. These cryptocurrencies act as a kind of collateral that supports transactions and network maintenance. As more users participate in staking, the network becomes more secure and efficient.

The rewards earned through staking may vary depending on the cryptocurrency and staking protocol used. Rewards are usually distributed proportionally to the number of cryptocurrencies locked and the time for which they remain blocked. These rewards can be in the form of more cryptocurrencies of the same type being blocked, which helps increase the amount of assets participants own.

Staking offers several benefits for both individual users and the network. Some of the benefits include:

1. Passive income generation: Staking allows cryptocurrency holders to earn additional rewards simply for keeping their assets locked. This provides an opportunity to earn passive income without the need to make active trades in the financial markets.

2. Contribution to network security: By participating in staking, users contribute assets that support the blockchain network and increase its security. By having a greater participation of users with locked cryptocurrencies, the risk of malicious attacks or manipulations is reduced.

3. Greater decentralization: Staking encourages greater participation and ownership of the network by users. This promotes greater decentralization and prevents network control from being concentrated in a few entities or miners with a lot of computing power.

Although staking can offer benefits, it also carries some risks and considerations. Some of them include:

1. Security risk: When blocking cryptocurrencies, there is a risk that the wallet or staking platform will be attacked or compromised, which could result in the loss of the blocked assets.

2. Liquidity risk: By locking cryptocurrencies, users cannot easily access those assets for use or sale. This can limit liquidity and financial flexibility.

3. Technical requirements: Participating in staking may require technical knowledge and experience in managing

specific portfolios and platforms. Users should be familiar with the appropriate requirements and procedures to avoid unnecessary risks.

In short, staking is a process by which cryptocurrency holders lock their assets to back a specific blockchain network and receive rewards in return. It provides benefits such as passive income, contribution to network security, and increased decentralization. However, it also carries risks and considerations that users should be aware of when engaging in this practice.

3 ALTCOINS Y TOKENS

What is the difference between these two?   Here is the explanation.

Altcoins and tokens are two important concepts in the crypto ecosystem that represent different types of digital assets. Understanding the difference between them is critical to understanding the diversification of the crypto ecosystem.

Altcoins, or "altcoins," are cryptocurrencies that were created after Bitcoin. Although they share similar characteristics to Bitcoin, such as using blockchain technology, they have their own distinctive characteristics. Some altcoins seek to improve specific aspects of Bitcoin, such as transaction speed, scalability or privacy. Examples of popular altcoins include Litecoin, Ethereum, Ripple, and Bitcoin Cash. Each altcoin has its own independent blockchain and unique set of features and use cases.

On the other hand, tokens are digital assets that exist on top of an already existing blockchain, such as Ethereum. Unlike altcoins, tokens do not have their own independent blockchain. Instead, they are created using specific token standards, such as the ERC-20 standard on Ethereum. Tokens are issued and transferred on the base blockchain and benefit from the security and features of the underlying network. The tokens are used to represent assets, rights, or utilities within specific projects, such as decentralized finance (DeFi) projects or governance. Some examples of tokens popular are : Tether (USDT), Binance Coin (BNB) y Chainlink (LINK).

The diversification of the crypto ecosystem refers to the wide variety of altcoins and tokens available in the market. As interest in cryptocurrencies has grown, numerous altcoins and tokens have been created to address different needs and use cases. This diversification offers investors and users the opportunity to explore and participate in projects with unique approaches and technologies. By diversifying their cryptocurrency portfolio, users can reduce risk and take advantage of opportunities in different sectors and projects in the crypto space.

Importantly, both altcoins and tokens have their own unique strengths, weaknesses, and characteristics. When considering diversification in the crypto ecosystem, it is essential to research and understand the fundamentals of each altcoin or token, as well as evaluate its technology, development team, use cases, and community. Smart diversification can help investors tap into growth potential and mitigate risks in the changing world of cryptocurrencies.

Basically, a token is created under an Altcoin and when you store your tokens in a cold wallet you must store them under your Altcoins.

For example, Shiba Inu you must save it in Etherium but by doing this they do not convert or exchange to Etherium are only stored there, and you can see them and save their value if it goes up or down.

It should also be noted that there are coins of different layers are called layers in English here you have a detailed summary of their difference and what they are for?

In the context of cryptocurrencies and blockchains, the term "layers" is used to describe different levels of functionality and complexity in the design and operation of projects. Each layer has its own purpose and specific characteristics. Next, I will explain the most common layers in the context of tokens and cryptocurrencies:

1. Layer 1:
Layer 1 or Layer 1 refers to the foundation of the infrastructure of a cryptocurrency or blockchain. In this layer, you define the fundamental protocol that establishes the basic rules of the network, including block creation, transaction validation, and security. Layer 1 is also responsible for the generation and distribution of the network's native cryptocurrency.

Examples of Layer 1 include Bitcoin, Ethereum and Cardano. Each of these cryptocurrencies has its own unique Layer 1 protocol and rules that determine how blocks are created and the network is secured.

2. Layer 2:
Layer 2 or Layer 2 refers to scalability solutions that are built on top of Layer 1 to Improve network performance and efficiency. These solutions seek to solve the challenges of

congestion and high transaction fees that can occur on Layer 1 blockchains.

Examples of Layer 2 include scalability solutions such as the Lightning Network for Bitcoin and sidechain solutions for Ethereum. These additional layers allow transactions to be made off-chain and then anchor the results to the mainchain, reducing the workload on Layer 1 and improving the speed and cost of transactions.

3. Layer 3 and more:
In some cases, Layer 3 and more can be mentioned, although these terms are not as common. Layer 3 generally refers to solutions or applications built on top of Layer 2. These applications can be more complex smart contracts, DApps (decentralized applications) or even additional protocols and platforms that build on existing Layer 1 and Layer 2 infrastructure.

Importantly, the use of "layers" in the context of cryptocurrencies and tokens can vary and is not always applicable in all projects. Some cryptocurrencies may have different architectures and not strictly follow this layered classification. However, in general, the idea of layers is used to describe different levels of functionality and complexity in the design of cryptocurrencies and blockchains.

Some currencies also sometimes change names.

The phenomenon of a cryptocurrency forking and creating a new currency with a slightly different name, such As in the case of Ethereum (ETH) and Ethereum Classic (ETC), it occurs due to what is known as a "fork of the chain" or "hard fork" in English. This situation can arise when there are significant disagreements within a cryptocurrency community regarding certain aspects of the protocol or consensus rules.

Below is the process behind creating a new coin through a fork in the chain:

1. Disagreement in the Community: Disagreements in the community of a cryptocurrency can arise due to technical issues, governance changes, changes in consensus rules, security incidents, or even decisions related to the development of the project. These disagreements can divide the community into different factions with divergent opinions on how to move forward.

2. Hard Fork: When the community cannot reach consensus on how to handle disagreement, a fork of the chain can occur. At this point, some community members may choose to follow the original rules and protocols (original chain) while others choose to follow a modified version of the chain (new chain).

3. Creation of the New Currency: The new chain, which follows a different direction from the original, may lead to the creation of a new cryptocurrency with a slightly different name. In the case of Ethereum, the fork resulted in two chains: Ethereum (ETH) and Ethereum Classic (ETC). Both chains share the same history of transactions up to a certain point, but from the fork, they follow different routes.

It is important to note that a fork in the chain does not always result in a new currency. In some cases, one of the chains may lose support and be abandoned, while the other continues to develop and grow. However, in situations where both networks are still active and have community support, it is common for them to be given different names to distinguish them.

Each chain can have its own characteristics, communities, and goals, leading to different levels of adoption and valuation of the resulting cryptocurrencies. The decision to follow one or the other chain is usually based on the individual beliefs, preferences, and goals of investors and participants in the crypto ecosystem.

There is also something that is known as Fork.

A "fork" in the context of cryptocurrencies refers to a fork in the blockchain, which can result in two or more different versions of the same cryptocurrency. This fork can occur due to updates, changes in consensus rules, disagreements in the community, or security incidents. Depending on the nature of the fork, they can be classified into two main types: hard forks and soft forks.

1. Hard Fork:
A hard fork occurs when a change is introduced in the protocol of a cryptocurrency that is incompatible with the previous version. This means that the new rules and features of the protocol are not backward compatible with the previous version. As a result, the blockchain IS divides into

two independent branches: the original version and the new version. Each of these branches continues with its own path and records transactions separately from the bifurcation point.

A famous example of a hard fork is the case of Ethereum and Ethereum Classic. In 2016, after a hacker attack on an Ethereum-based project called the DAO, the Ethereum community could not agree on how to handle the situation. This resulted in a fork of the chain, where most of the community opted to follow an updated version of the protocol, thus creating Ethereum (ETH). Meanwhile, a minority group decided to keep the original chain unchanged and called it Ethereum Classic (ETC).

2. Soft Fork:
A soft fork occurs when a change is introduced in the protocol that is backward compatible. That is, the new rules and features of the protocol are backward compatible with the previous version. In a soft fork, only nodes that update their software comply with the new rules, while nodes that continue to use the old version can still participate in the network. This means that the blockchain is not divided into separate branches, and all participants remain on the same chain.

In short, a fork in a coin refers to a fork in the blockchain that can divide the community and give rise to two or more versions of the same cryptocurrency. Forks are major events in the crypto ecosystem and can have significant implications. in terms of governance, security and currency adoption.

4 MY KEYS MY COINS

Let's explain this concept a little:

Each cryptocurrency, Altcoin or Token has a unique number of bit keys here is the most in-depth explanation.

The number of digits in a cryptocurrency's key can vary depending on the cryptographic algorithm used and the security standard implemented. In general, the cryptographic keys used in cryptocurrencies usually have a fixed length expressed in bits.

For example, Bitcoin uses the asymmetric cryptography algorithm (public key/private key) known as curved elliptic (ECDSA) with a key length of 256 bits for addresses and signatures. This translates to a 256-bit private key and a 512-bit public key.

Other cryptographic protocols may use different key lengths. For example, Ethereum also uses the elliptic curve algorithm, but with a key length of 160 bits for addresses.

It is important to note that the length of the key is not expressed in terms of decimal digits, but in bits. One bit can take the value of 0 or 1, and 8 bits equals 1 byte. Thus, a Bitcoin private key, for example, consists of a sequence of 256 bits, which can be expressed as 32 bytes or 64 hexadecimal digits.

In summary, the cryptographic keys used in cryptocurrencies have a fixed length expressed in bits, and the number of digits depends on the representation system used (for example, hexadecimal). Keys are typically 256 bits or more long, depending on the cryptographic algorithm and the level of security required.

When you keep your coins you do not keep the coins in itself what you actually keep are your private keys which gives you control of them to only you possess that private key there is no one who can steal them sit or you do not give access is said out there: my keys my Cryptocurrencies if you do not have the keys then you can say that the coins are not yours

The concept "my keys, my cryptocurrencies" refers to the idea that control of the private keys of a cryptocurrency is essential to having control and ownership of those cryptocurrencies. In the context of cryptocurrencies, a private key is a cryptographically generated sequence of characters that acts as a password or key to access and control funds from a specific cryptocurrency address.

When talking about "my keys, my cryptocurrencies", the importance of users having exclusive control of their private keys is emphasized. This means that private keys should be kept in a safe and trusted place, preferably offline and out of the reach of unauthorized third parties. If a user loses or has their private keys stolen, they will also lose access to their cryptocurrencies, as they will not be able to prove their ownership and authority over them.

The idea behind this concept is that, by having sole control of private keys, users can maintain the security of their cryptocurrencies and avoid relying on intermediaries or third parties to access and manage their digital assets. This also avoids the risk of private keys being vulnerable to hacks or security breaches on exchange platforms or online wallets.

A common way to follow the principle "my keys, my cryptocurrencies" is to use cryptocurrency wallets that give users full control of their private keys. These wallets, such as hardware wallets or software wallets that enable offline private key generation and storage, give users greater security and control over their digital assets.

Importantly, maintaining exclusive control of private keys is essential to having full control of private keys. Cryptocurrencies also come with greater responsibility. Users should ensure they have adequate backups of their private keys and follow security best practices to protect their sensitive information.

There are different types of wallets where you can store your coins either online or offline.
You can also leave your coins on the exchange that you buy them.
It is necessary to explain that if you buy crypto in an exchange you can leave them there but there is eel risk that the exchange is fenced to bankruptcy and you no longer have access to your coins and lose them There is also the risk that the exchange does not allow you to sell your coins because it is going through a moment of internal problems and it is not convenient for them to sell them and you can

Cryptocurrency the digital monetary revolution

only do it when they allow you pore so is better : my keys my coins

5 TYPES OF WALLETS

Here we will see the types of wallets to store your coins.

There are different types of wallets for holding cryptocurrencies, each with its own security and convenience features. Next, I will explain the main types of crypto wallets:

1. Software wallets: Also known as desktop wallets or mobile wallets, these are downloadable apps on your computer or mobile device. You can install these wallets on your device and have access to your cryptocurrencies at any time. Some popular software wallets include Exodus, Electrum, Trust Wallet, and Atomic Wallet. These wallets are convenient and easy to use, but you should make sure to keep your operating system up to date and use additional security measures, such as passwords and two-factor authentication.

2. Hardware wallets: Hardware wallets are physical devices specially designed to store your cryptocurrencies securely. These wallets store the private keys offline, making them less vulnerable to cyberattacks. Popular examples of hardware wallets are Ledger Nano S, Trezor, and KeepKey. To use a hardware wallet, you'll need to connect it to your computer or mobile device when you want to make a transaction.

3. Hot wallets: Online wallets, also known as hot wallets, are cloud-based wallets that allow you to access your cryptocurrencies through a website or an online application. These wallets are convenient and easy to use, as you can

access them from any device with an internet connection. However, being online, they are more exposed to potential security risks, such as hacking or phishing. Some popular online wallets include Coinbase Wallet, Binance Wallet, and MetaMask.

4. Paper wallets: A paper wallet is a physical way to store your private keys on paper or another physical medium. You can generate a private key offline, print it and store it in a safe place. These wallets are safe against online attacks, as they are not connected to the Internet, but You must make sure to protect the paper or physical medium from damage or loss. Paper wallets are especially useful as a long-term storage option or for storing large amounts of cryptocurrency securely.

It's important to remember that regardless of the type of wallet you choose, you should take extra security precautions, such as keeping your devices up to date, using strong passwords, turning on two-factor authentication, and regularly backing up your private keys. It is also critical to research and choose trusted wallets from reputable developers and verify the authenticity of apps or devices before using them.

Some exchanges offer services that allow users to store their cryptocurrencies directly on the platform. This implies that the coins are kept at an exchange address and are accessible through the user's account on the platform. However, you should keep in mind that leaving your coins on an exchange involves some risks and considerations.

1. Security: By leaving your coins on an exchange, you are entrusting the custody of your assets to a third party. This means that the security of your coins is in the hands of the platform and is subject to risks such as hacks, cyber-attacks or other security vulnerabilities. It is important to choose reputable exchanges with strong security measures and two-factor authentication to minimize these risks.

2. Control: By using an exchange as a storage location, you don't have full control over your private keys. Private keys are managed by the exchange on your behalf, which means you can't directly access them or fully control your assets. In case of problems with the exchange, you could face difficulties accessing your coins or recovering your funds.

3. Revenue Generation: Some exchanges offer revenue generation programs for the cryptocurrencies that are held on their platform. This is known as "staking" or "rewards programs." By participating in these programs, you can earn Interest, rewards or dividends on your cryptocurrencies. The exchange uses your coins to support its operation and, in return, offers you a portion of the revenue generated. This can be a way to generate passive income with your cryptocurrencies while keeping them on the exchange.

It is important to research and understand the terms and conditions of the income generating programs offered by exchanges. Some important considerations include the length of the revenue generation period, the interest rates or rewards offered, and the risks associated with the platform. You should also consider whether income generation justifies the risks and loss of control over your assets.

In short, storing your coins on an exchange can offer convenience and the ability to participate in income-generating programs. However, it also involves security risks and loss of control over your private keys and assets. It's important to research and choose reliable exchanges and understand the associated risks before deciding on how to store your cryptocurrencies.

6 NFTS CREATIONS THROUGH SOME BLOCKCHAINS SUCH AS SOLANA

When creating altcoins and tokens, a new concept called NFTS emerged.

The translation of NFT into Spanish is "Token No Fungible". "Fungible" refers to the property of being interchangeable for identical units in terms of value and characteristics. Therefore, a "Non-Fungible Token" is a unique and indivisible type of digital asset that cannot be exchanged one by one for another token. Each NFT has its own identity and distinctive features, making it unique and non-exchangeable with other tokens.

Here are some examples of popular NFTs:

1.   "CryptoPunks": They are a series of 10,000 pixelated characters unique to the Ethereum blockchain. Each has individual characteristics and traits, making them coveted and collectible.

CryptoPunks are considered one of the first non-fungible token (NFT) projects in the history of cryptocurrencies and digital art. They were created by Matt Hall and John Watkinson, who are part of the software development studio Larva Labs, and launched in June 2017 on the Ethereum network.

Characteristics and Concept of CryptoPunks:

1. NFTs on the Ethereum Blockchain: CryptoPunks are non-fungible tokens based on the Ethereum blockchain. Each CryptoPunk is unique and cannot be replaced by another, which means that each has a unique value in the market.

2. Pixelated Art and Retro Style: CryptoPunks are designed as 8-bit pixelated characters, inspired by the graphic style of early video games from the 1980s. There are a total of 10,000 different CryptoPunks, each with a unique combination of traits such as hairstyles, eyes, hats, accessories, among others.

3. Initial Free Distribution: In the beginning, the 10,000 CryptoPunks were available for free, and anyone could claim them from the Larva Labs website. Users only needed to pay the Ethereum network transaction fee to get their CryptoPunk. At that time, many did not attach much importance to it, and only a few claimed it.

4. Scarcity and Secondary Market: As the popularity of CryptoPunks grew and demand increased, the scarcity of these unique characters became apparent. Owners began selling and trading them on the secondary market, and their prices skyrocketed due to their rarity and the value perceived by collectors.

5. Artistic and Speculative Value: CryptoPunks are considered a digital work of art and have become a symbol of crypto culture. Some collectors are willing to pay large sums of money for rarer and more unique CryptoPunks, which has led to significant growth in the NFTs market.

6. Inspiration for the Rise of NFTs: CryptoPunks are considered pioneers in the NFT space and have inspired many other digital art projects and collectibles in the crypto world. Their success has paved the way for the subsequent boom of NFTs and has opened the doors to the creation and commercialization of unique digital assets in various industries.

In short, CryptoPunks are a unique and valuable collection of non-fungible tokens that have left a significant mark on the history of cryptocurrencies and digital art. Their popularity has demonstrated the potential and importance of NFTs as unique digital assets, and their legacy continues to influence the development of new projects and applications in the crypto space.

2. "Bored Ape Yacht Club": It is a collection of 10,000 computer-generated monkey images on the Ethereum blockchain. Each monkey has different characteristics and belongs to an exclusive owner.

3. "Art Blocks": It is a platform on Ethereum that allows artists to create generative art. Collectors can acquire unique and limited generative art pieces.

4. "Cryptokitties": It is one of the first NFT-based games on the Ethereum blockchain. Users can breed and collect unique virtual cats with varying genetic characteristics.

5. "NBA Top Shot": It is a platform that offers NBA highlights as NFTs. Users can collect, buy and sell unique video moments of famous NBA plays.

6. "Decentraland": It is a decentralized virtual reality platform based on the Ethereum blockchain. Users can buy, sell and own virtual land, as well as unique objects and avatars such as NFTs.

These are just a few examples of popular NFTs, but the world of NFTs is very broad and diverse, with a wide range of digital collectibles, art, games and virtual experiences being created on different blockchains.

To create NFTs (Non-Fungible Tokens) on a specific blockchain, such as Solana, it is necessary to follow certain steps and use the right tools. Below, I will provide you with a general explanation of how NFTs can be created using altcoins, such as Solana, as an example:

1. Choose a supported platform or protocol: To create NFTs on Solana, you will need to use a platform or protocol compatible with the Solana blockchain that supports the creation and issuance of non-fungible tokens. Some examples of popular platforms for creating NFTs in Solana are Solible, Digital Eyes, and Solana Art.

2. Account and wallet configuration: To get started, you will need to set up an account on the chosen platform and have a Solana-compatible wallet that allows you to interact with the blockchain. You can use wallets such as Sollet, Phantom or Solflare, among others.

3. NFT creation and customization: Once you've set up your account and wallet, you'll be able to create and customize your NFT. This includes providing details such as name,

description, image or file associated with the NFT, as well as other relevant metadata.

4. Owner assignment and issuance: After customizing the NFT, you will need to assign an owner to the token. This can be your own account or someone else's account you want to transfer the NFT to. Once you have assigned the owner, you will be able to issue the NFT on the Solana blockchain, which will make it unique and verifiable.

5. Listing and trading: Once you have created and issued the NFT, you will have the option to list and trade the token on markets and platforms specialized in NFTs. These markets will allow you to list or auction your NFT, and other users will be able to buy it using cryptocurrencies, including altcoins.

It is important to note that the specific process for creating NFTs may vary depending on the platform or protocol you choose, as well as the features and options available in each. In addition, it is essential to research and understand the fees, requirements and considerations of each platform and blockchain before creating and marketing NFTs.

In short, to create NFTs using altcoins like Solana, it is necessary to choose a supported platform, set up an account and wallet, customize the NFT, assign an owner, issue the NFT on the blockchain, and finally list and trade the token on specialized marketplaces.

NFTs (Non-Fungible Tokens) in Solana are created using smart contract technology and standards that offers the

Solana network. The following explains the general process for creating NFTs in Solana:

1. Smart Contract: In Solana, NFTs are created by implementing a specific smart contract that defines the unique characteristics and properties of the NFT. Smart contracts in Solana are developed using the Rust programming language.

2. NFT Metadata: Each NFT contains metadata that describes its attributes and characteristics. This metadata may include information such as the name of the NFT, a description, the creator, an image or link to an image representing the NFT, and any other relevant information.

3. Standard Token: In order for the NFT to be compatible with the Solana network and other projects that use it, certain standards are followed. One of the most widely used standards for NFTs in Solana is the SPL (Solana Program Library) standard, which defines how NFTs should be structured and handled on the blockchain.

4. Minting: The process of creating an NFT is known as "minting". When a new NFT is created, the metadata and unique characteristics are assigned to the token and associated with a specific address in the Solana network.

The process of "minting" refers to the creation of new tokens, whether cryptocurrencies or NFTs (Non-Fungible Tokens), on a specific blockchain or blockchain platform. Minting is an essential function in blockchains that enable the creation of new digital assets and give them value and

Authenticity in the crypto ecosystem. The minting process and its purpose are explained below:

1. Creation of New Tokens: Minting is used to generate new tokens on a blockchain. For example, if a project decides to launch its own cryptocurrency or NFT, it will need to go through the minting process to create and allocate those new tokens on the blockchain.

2. Smart Contracts: On many blockchain platforms, such as Ethereum or Solana, new tokens are created using smart contracts. These smart contracts are programs that contain rules and logic for the creation and management of tokens.

3. Property Mapping: During the minting process, unique properties and characteristics can be assigned to new tokens. For example, in the case of NFTs, metadata can be added that describes the artwork, music, or any other digital asset that the NFT represents.

4. Authenticity and Scarcity: One of the main purposes of minting is to ensure the authenticity and scarcity of tokens. For example, in the case of NFTs, each token has its own unique identifier, ensuring that it is a unique and irreplaceable digital asset.

5. Monetization and Economy: The minting process allows token creators or issuers to establish business models and economies around their digital assets. For example, NFT creators can sell their works from digital art at auctions or marketplaces, allowing them to monetize their art and establish market value.

6. Tokens and Cryptocurrencies: In addition to NFTs, minting also applies to the creation of new cryptocurrencies and fungible tokens. The new cryptocurrencies can be used for various applications, such as means of payment, incentives or governance in a blockchain network.

In short, minting is an essential process for the creation of new tokens and digital assets on a blockchain. It allows issuers to create unique and authentic assets, contributing to diversity and innovation in the crypto ecosystem. In addition, minting plays a pivotal role in creating economies and business models around digital assets, providing opportunities for creators, investors, and enthusiasts of blockchain technology.

5. Wallets: NFTs created in Solana are stored in network-compatible wallets, such as Sollet or Phantom. Users can view, transfer, and manage their NFTs through these wallets.

6. Auctions and Markets: Once created, NFTs can be listed on marketplaces and platforms specialized in the sale and purchase of NFTs. Collectors and investors can bid or buy NFTs according to their preferences.

Importantly, Solana has gained popularity in the creation and trading of NFTs due to its scalability and low transaction costs. However, before creating or buying NFTs on Solana, it is critical to verify the authenticity of the smart contract and the reputation of the platforms and marketplaces that offer them. As always, it is It is advisable to investigate and take precautions to avoid possible scams or risks.

7 CRYPTOGRAPHIC CREDIT CARDS

So, there are already crypto credit cards here we will see some of them:
"Crypto credit cards" are payment cards that allow users to spend their cryptocurrencies directly in establishments that accept card payments. These cards work similarly to traditional credit or debit cards, but instead of using fiat currency, they are linked to a user's crypto wallet.

When a user makes a transaction with a crypto credit card, the card automatically converts the selected cryptocurrency into fiat currency in real time. This allows the merchant to receive payment in traditional currency, while the user spends their cryptocurrencies without having to convert them previously.

Crypto credit cards give users the convenience of using their cryptocurrencies in everyday life and in places where cryptocurrencies are not yet widely accepted. Some of these cards offer additional benefits, such as cryptocurrency rewards, discounts, and loyalty programs.

Importantly, crypto credit cards are associated with certain costs and considerations. There may be issuance and usage fees, as well as conversion fees from cryptocurrency to fiat currency. In addition, each card has its own security and data protection policies that must be taken into account.

When choosing a crypto credit card, it is critical to research and consider several factors, such as supported cryptocurrencies, fees, spending limits, geographic availability, and the reputation of the card issuing company.

In short, crypto credit cards are payment cards that allow users to spend their cryptocurrencies in establishments that accept card payments. These cards automatically convert cryptocurrency into fiat currency in real time, making it easier to use cryptocurrencies in everyday life. However, it is important to research and consider the associated costs and policies before choosing a specific card.

Sofi Credit Card is a credit card offered by SoFi, an online financial services company. This card is designed to provide users with a comprehensive financial experience, combining traditional credit card benefits with the option to earn rewards in cryptocurrencies.

One of the distinguishing features of the SoFi Credit Card is its "Crypto Rewards". With this feature, users can earn cryptocurrencies instead of traditional points or miles when making purchases with their card.

Below are some key details about SoFi Credit Card Crypto Rewards:

1. Supported cryptocurrencies: SoFi allows users to earn Bitcoin (BTC) or other selected cryptocurrencies as a reward for their purchases. Available cryptocurrencies may vary depending on the company's current offering.

2. Percentage of rewards: Users can earn a specific percentage in cryptocurrencies for every dollar spent with the card. The percentage of rewards can vary and is usually shown as a number followed by "sats", which is short for "satoshis", the smallest unit of Bitcoin.

3. Accumulation and redemption: Cryptocurrency rewards are automatically accumulated in the user's account as they make eligible purchases. These cryptocurrencies can be held as a long-term investment or can be sold or redeemed based on the user's preference.

4. Additional benefits: In addition to Crypto Rewards, SoFi Credit Card may offer other standard credit card benefits, such as car rental insurance, purchase protection, and travel assistance.

It is important to note that crypto rewards are subject to specific terms and conditions set by SoFi. Each credit card may have different levels of rewards, limits, and policies associated with it. Therefore, it is advisable to review and understand in detail the details of the SoFi Credit Card and its Crypto Rewards before applying for or using it.

. Brex credit card BREX is a financial services company that offers a credit card specially designed for businesses. The Brex Credit Card for Startups is aimed at startups and emerging companies, providing a variety of benefits and rewards, including Crypto Rewards.

The Brex Credit Card for Startups allows businesses to earn cryptocurrency rewards for their business expenses. Below are some key aspects about Brex Crypto Rewards:

1. Supported Cryptocurrencies: Brex offers the option to earn Bitcoin (BTC) or other selected cryptocurrencies as a reward for trading expenses. The specific selection of cryptocurrencies may vary and is subject to the company's current offering.

2. Rewards Percentage: Brex card users can earn a specific percentage in cryptocurrency for every dollar spent in eligible merchant categories. The percentage of rewards may vary and is set according to Brex's terms and conditions.

3. Accumulation and redemption: Cryptocurrency rewards are automatically accrued to the company's account as eligible trading purchases are made. These cryptocurrencies can be held as a long-term investment or can be sold or redeemed depending on the company's preference.

4. Additional Benefits: In addition to Crypto Rewards, the Brex Credit Card for Startups offers other business benefits, such as high credit limits, simplified business spending policies, and online financial management tools.

It is important to note that the specific features and benefits of the Brex card may vary depending on the company's current offer and are subject to specific terms and conditions set by Brex. It is recommended to review and understand in detail the details of the Brex Credit Card for Startups, including Crypto Rewards, before applying for or using it.

Crypto.com Visa Card is a debit card offered by the Crypto.com platform. This card allows users to spend their cryptocurrencies in establishments that accept card payments, providing a convenient way to use digital assets in everyday life. In addition, the card offers cryptocurrency rewards to cardholders.

Here are key details about the Crypto.com Visa Card and its features:

1. Supported cryptocurrencies: The Crypto.com Visa Card supports a wide range of cryptocurrencies, including Bitcoin (BTC), Ethereum (ETH), Litecoin (LTC) and many others. Users can load their cryptocurrencies onto the card and use them to make purchases at establishments that accept Visa.

2. Percentage of rewards: The card offers rewards in cryptocurrencies for each transaction made with the card. The percentage of rewards varies depending on the type of card and the level of staking required (if applicable). Cardholders can earn up to a certain percentage in cryptocurrency for every purchase they make.

3. Staking Requirement: To access certain levels of rewards and additional benefits, cardholders may need to meet a staking requirement. Staking involves blocking a certain Number of cryptocurrencies on the Crypto.com platform for a certain period. By complying with this requirement, the

Users can gain access to better rewards and additional benefits.

4. Additional benefits: In addition to cryptocurrency rewards, the Visa Card Crypto.com offers other benefits such as cash back, access to premium services, loyalty programs and discounts on partner services and merchants.

It is important to note that the specific details of the Visa Card Crypto.com and its features may vary depending on the country of residence and the card level selected. Therefore, it is recommended to review the terms and conditions and conduct a detailed investigation before applying for and using the card.

The Coinbase Card is a debit card offered by Coinbase, a renowned cryptocurrency exchange and wallet. This card is designed to allow users to spend their own cryptocurrencies in establishments that accept card payments, providing a convenient way to use digital assets in everyday transactions.

Below are some key details about the Coinbase Card and its features:

1. Supported cryptocurrencies: The Coinbase Card supports several popular cryptocurrencies, such as Bitcoin (BTC), Ethereum (ETH), Litecoin (LTC) and many more. Users can load their card accounts with their cryptocurrencies y Use them to make purchases in physical establishments or online.

2. Percentage of rewards: The card offers rewards in cryptocurrencies for each transaction made with the card. The percentage of rewards may vary depending on the current rewards program and the user's geographic location.

3. Ease of use: The Coinbase Card is easy to use as it can be used like any other traditional debit card. Users can make payments at any establishment that accepts Visa card payments.

4. Integration with Coinbase wallet: The card is linked to the user's Coinbase wallet, allowing for seamless integration to transact with cryptocurrencies and manage the balances of the associated cryptocurrencies.

5. Management and Control: Users can monitor and manage their transactions, check cryptocurrency balances, and receive real-time notifications through the Coinbase mobile app.

It is important to note that the specific details of the Coinbase Card, such as reward percentages and supported cryptocurrencies, may vary depending on geographic location and current rewards programs. Therefore, it is recommended to review the specific terms and conditions of the card and conduct detailed research before applying for and using it.

8 THE "MEME" COINS

Now let's see the MEME coins those currencies that create rich overnight and do not have much strength in social networks causing the rise in their price.

Meme coins, also known as "cryptocurrency memes", are a type of cryptocurrency that are characterized by their humorous or ironic nature, and are often based on memes, internet topics or popular phenomena on social networks. These cryptocurrencies typically have limited or no utility beyond their playful aspect, and their value and popularity are often driven by online communities and social media.

Next, I explain in more detail the characteristics and key aspects of meme coins:

1. **Origin and Popularity:** Meme coins gained popularity in the cryptocurrency world thanks to the growing influence of social media, especially on platforms such as Reddit, Twitter, and TikTok. These online communities have fun creating and sharing memes about cryptocurrencies, and sometimes create their own cryptocurrencies based on those memes as a kind of joke or parody.

2. **Ironic and Humorous Nature: ** The main characteristic of meme coins is their ironic and humorous nature. They often poke fun at aspects of cryptocurrency culture and investors, and often have names and symbols related to popular memes or viral topics on the internet.

3. **Limited Utility: ** In many cases, meme coins have limited or no utility. Unlike other cryptocurrencies that are used to transact, execute smart contracts, or provide specific services, meme coins typically lack a practical, real-world use case.

4. **Volatility and Speculation: ** Due to their playful nature and lack of a specific purpose beyond entertainment, meme coins tend to be extremely volatile in terms of price. Their value can increase rapidly due to speculation and social media activity, but they can also lose value quickly.

5. **Risks and Precautions: ** Investing in meme coins carries significant risks, as their prices can be highly influenced by speculation and social media trends. Investors should keep in mind that investing in meme coins can be extremely risky and that they could lose their money if the value of the cryptocurrency suddenly plummets.

Some examples of popular meme coins include Dogecoin (based on the Shiba Inu dog meme), Shiba Inu (a fork of Dogecoin), and SafeMoon (with a mechanism of rewards and penalties for holders).

In conclusion, meme coins are cryptocurrencies based on memes and internet themes, with an ironic and humorous nature. While they can be fun and attract attention on social media, it's important that investors understand the risks involved and stay informed before investing in them.

Let's now see some examples of them.

Shiba Inu (SHIB) is a cryptocurrency that belongs to the category of meme coins, inspired by the famous Shiba Inu dog meme that went viral on the internet. It was created in August 2020 by an anonymous developer under the pseudonym "Ryoshi". This cryptocurrency has gained popularity due to its playful and humorous nature, as well as its active community on social media.

Here are some key aspects of Shiba Inu:

1. **Origin and Design: ** Shiba Inu was designed as an homage to the Shiba Inu dog meme, which became famous as the symbol of the Dogecoin currency. Its design and name are inspired by this breed of dogs and refer to the meme culture of the internet.

2. **Supply and Tokenomics: ** Shiba Inu was created as a token on the Ethereum network and uses the ERC-20 standard. The total supply of SHIB is extremely large, with a quadrillion (10^{15}) of initial tokens, making each token extremely low in value. This feature is part of the humorous and playful approach of the coin.

3. **Active Community: ** Shiba Inu has developed an active and passionate community on social media, especially on Twitter and Reddit. Community members share memes,

discuss project development and promote the adoption of SHIB.

4. **DeFi and Token Burning: ** Shiba Inu has attempted to take advantage of the growth of the DeFi (decentralized finance) ecosystem by launching ShibaSwap, a decentralized exchange platform. They have also implemented a token burning mechanism, in which a percentage of each transaction is permanently removed from the supply, which could influence the value of the remaining tokens.

5. **Risks and Speculation: ** As with other meme coins, Shiba Inu is highly speculative and volatile in terms of price. Investors should exercise caution and be willing to take considerable risks, as their value can change drastically due to speculative and market factors.

Importantly, although Shiba Inu has gained popularity and attracted many investors, its playful nature and lack of a meaningful practical use case make it a high-risk and speculative investment. As with any cryptocurrency investment, it is recommended to conduct thorough research and be informed about the risks before investing in Shiba Inu or other meme coins.

Now let's see PEPE coin :

Created in April 2023 it has been an exposure on social networks and is accepted by many who are affected by the FOMO "FEAR OF BEING LEFT OUT OF THE GAME"

Let's see now DOGE coin:

Dogecoin (DOGE) was originally created as a joke and parody of cryptocurrencies in December 2013 by two software developers, Billy Markus and Jackson Palmer. Its creation was based on the popular internet meme "Doge", which depicts a Shiba Inu breed dog accompanied by poorly written English phrases in a comical style. The choice of the name and logo of Dogecoin was inspired by this meme, which gave it a humorous and playful approach from its conception.

Despite starting as a joke, Dogecoin quickly gained popularity and an active online community. Followers of the Doge meme loved the idea of bringing the fun and joyful spirit of the meme to a real cryptocurrency. In addition, the fact that Dogecoin had a virtually unlimited initial supply (100 billion DOGE was mined in the first year) and very low transaction fees, made it attractive to send small tips and donations online.

Over the years, the Dogecoin community has been involved in various charity campaigns and fundraising events, including funding to send the Jamaican bobsleigh team to the Olympic Games in Winter de Sochi in 2014 and Donations for humanitarian and disaster relief projects.

Dogecoin's popularity soared further in early 2021, when endorsements from public figures like Tesla and SpaceX CEO Elon Musk and other social media comments boosted its value and visibility. However, it is important to note that due to its humorous and playful nature, Dogecoin remains highly speculative, and its price can experience large fluctuations based on factors such as social media trends and comments from public figures.

Despite being considered a meme coin, Dogecoin has proven to be more than just a joke, as it has found a loyal community and has been adopted for practical use on several occasions. However, investors and users should be aware of the risks associated with investing in any cryptocurrency, including Dogecoin, and should always conduct proper research before making financial decisions.

In summary there is a great variety of coins and tokens that grows every day more and more because day by day new coins continue to come to market.

9 COIN SWAP

There are several places where you can exchange currencies such as Pancake swap and Uniswap. Let's see some characteristics of them.

PancakeSwap is a decentralized exchange protocol (DEX) that operates on the Binance Smart Chain (BSC) blockchain network. It was launched in September 2020 and has become very popular in the world of decentralized finance (DeFi) due to its low transaction costs and speed compared to other protocols operating on the Ethereum network.

Here are the key aspects of PancakeSwap:

1. **Decentralized Exchange Protocol: ** PancakeSwap is a DEX, which means that it allows users to trade cryptocurrencies directly with each other without the need for intermediaries such as centralized exchanges. DEXs run on blockchain and use smart contracts to facilitate exchanges securely and transparently.

2. **Liquidity and Liquidity Pools: ** In order for users to be able to trade their cryptocurrencies on PancakeSwap, the protocol must have sufficient liquidity on each exchange pair. This is achieved by creating liquidity pools. Users can provide funds to these pools and, in return, receive LP tokens (liquidity tokens) that represent their participation in the pool. Through these pools, users can earn fees for providing liquidity to the protocol.

3. **Farm and Staking:** PancakeSwap offers farm and staking features that allow users to earn additional rewards in the form of native protocol tokens (CAKE) for participating in certain activities. Users can deposit their tokens in farms and staking pools to receive additional rewards in CAKE.

4. **Binance Smart Chain Fees and Benefits: ** PancakeSwap operates on Binance Smart Chain, a sidechain of the Binance Chain network. This network offers low transaction costs and higher processing speed compared to Ethereum, making PancakeSwap an attractive option for those looking to avoid high gas fees and congestion on the Ethereum network.

5. **Security and Smart Contracts: ** Although PancakeSwap is decentralized, the smart contracts that support it still present risks. Therefore, it is critical that users exercise caution when interacting with the protocol and make sure to carefully review the addresses and contracts they use.

PancakeSwap has gained popularity for its ease of use, low transaction costs, and wide selection of cryptocurrencies available to exchange. However, as with any DeFi protocol, it is always recommended that users conduct thorough research and understand the associated risks before engaging in any decentralized financial activity.

Farming in PancakeSwap is a strategy of participation in the decentralized exchange protocol (DEX) of PancakeSwap to earn additional rewards in the form of CAKE tokens (the native token of PancakeSwap). This strategy is known as "yield farming" or "liquidity farming" in the world of decentralized finance (DeFi).

Here's how farming works at PancakeSwap:

1. **Liquidity Pools:** In order for users to trade their cryptocurrencies on PancakeSwap, the protocol requires sufficient liquidity on each exchange pair. This liquidity is achieved through the creation of liquidity pools. Users can provide funds to these pools, depositing a pair of specific cryptocurrencies, and in return, they receive LP tokens (liquidity tokens) that represent their participation in the pool.

2. **Farming with LP:** Once users have LP tokens from a specific pool, they can use them to participate in farming. This involves locking LP tokens on certain farms designated by PancakeSwap. These farms are smart contracts that allow users to "grow" or "farm" rewards in the form of CAKE tokens.

3. **Rewards in CAKE:** Users who participate in farming with their LP tokens get rewards in the form of CAKE tokens. These rewards are generated thanks to the fees charged in the protocol for each exchange and are distributed among the liquidity providers participating in the farming.

4. **Performance and Risks:** Farming on PancakeSwap offers users the opportunity to earn additional return in the form of CAKE tokens. However, it also carries certain risks. The volatility of the mercado, fluctuations in cryptocurrency prices and changes in demand can affect the value of rewards in CAKE and LP tokens.

5. **Staking and Unstaking:** Users can participate in farming by depositing their LP tokens at the corresponding farms and earn rewards continuously. They also have the flexibility to withdraw their LP tokens (unstaking) at any time. However, it is important to note that some liquidity pools have lock-up periods during which LP tokens cannot be withdrawn.

Importantly, farming in PancakeSwap, as in other DeFi protocols, presents risks and challenges. Users should fully research and understand how farming works before engaging and consider the associated risks. It is also important to use secure wallets and follow security best practices when operating on DeFi.

Here we go with Uniswap.

Uniswap is a decentralized exchange protocol (DEX) that operates on the Ethereum blockchain network. It was launched in November 2018 by Hayden Adams and has become one of the most popular and used DEXs in the decentralized finance (DeFi) ecosystem.

Here are the key aspects of Uniswap:

1. **Decentralized Exchange:** Uniswap allows users to trade cryptocurrencies directly with each other without the need for an exchange or intermediary. Exchanges are carried out in a decentralized way through smart contracts on the Ethereum network. This means that users always have full

control of their funds, and it is not necessary to rely on a centralized entity to manage the exchanges.

2. **Liquidity and Liquidity Pools:** In order for trades to be conducted efficiently, Uniswap uses liquidity pools containing funds in both cryptocurrencies of the exchange pair. Users can contribute funds to these pools and in return receive liquidity tokens (LPs) that represent their participation in the pool. Liquidity providers earn fees for each trade made using pool funds.

3. **Automation and AMM:** Uniswap uses an automated market model (AMM), which means that there is no traditional order book as in centralized exchanges. Instead, prices are set automatically through algorithms based on supply and demand in liquidity pools. This ensures continuous liquidity and the ability to make trades quickly and without relying on the availability of counterparties.

4. **UNI Token:** Uniswap launched its own native token called UNI in September 2020. UNI holders have the right to participate in the governance of the protocol, allowing them to propose and vote on changes to the operation of Uniswap.

5. **Expansion and Versatility:** Uniswap has been a pioneer in the DeFi space and has been used for a variety of applications beyond cryptocurrency exchange, including lending, betting, yield farming and more. In addition, its success has inspired the development of other similar DEX protocols on different blockchains.

It is important to note that, as with any DeFi protocol, Uniswap presents risks and challenges, including those related to security and market volatility. Users should be careful when interacting with smart contracts and make sure to use secure wallets and follow security best practices when operating in the DeFi ecosystem.

10 THIS IS NOT AN INVESTMENT RECOMMENDATION

So, begin all the videos or explanations to explain some currency in if here what we are going to see is how to buy and analyze a cryptocurrency

Analyzing which cryptocurrency to buy involves conducting thorough research and detailed analysis, as the cryptocurrency market is highly volatile and presents significant risks. Here are some important steps and factors to consider when performing an analysis to make informed decisions about which cryptocurrency to buy:

1. **Project Research:** Start by thoroughly investigating the project behind cryptocurrency. It looks for information about its development team, its underlying technology, the purpose of the coin, and its use case. Examine the whitepaper and other whitepapers to understand how cryptocurrency works and what problems it tries to solve in the real world.

2. **Adoption and Practical Use:** Consider whether cryptocurrency has real adoption and whether it is used in real-world applications or platforms. Look for examples of use cases and companies that are using it in their operations.

3. **Liquidity and Trading Volume:** Check the liquidity and trading volume of the cryptocurrency in exchange houses. A cryptocurrency with a high trading volume and sufficient liquidity is usually more stable and easier to buy and sell.

4. **Trends and Market News:** Stay updated on the latest news and trends in the cryptocurrency market. News and events can significantly affect the price and demand for a particular cryptocurrency.

5. **Technical Analysis:** Consider performing technical analysis using charts and tools to identify price patterns and historical trends. This can help you make decisions based on the past behavior of the cryptocurrency price.

6. **Fundamental Analysis:** In addition to technical analysis, consider fundamental analysis, which involves evaluating economic, financial and market factors that may affect the price of cryptocurrency in the long term.

7. **Risks and Diversification:** Recognizes and understands the risks associated with investing in cryptocurrencies. The market is highly volatile and prices can change quickly. Consider diversifying your investments into different cryptocurrencies to reduce risk and maintain a long-term focus.

8. **Professional Advice:** If you have no

experience in cryptocurrency investments or analysis, consider seeking professional advice from financial experts or cryptocurrency specialists before making investment decisions.

Remember that no investment is risk-free, and the cryptocurrency market can be especially volatile. It is important to invest cautiously and only with funds that you can afford to lose. Always do your own research and make informed decisions based on your financial goals and risk tolerance.

You should also consider volatility because volatility in cryptocurrencies refers to the magnitude of the price changes that these digital currencies experience in a given period of time. In other words, it is the measure of how quickly and widely the price of a cryptocurrency can rise or fall in a short period.

Volatility is a common feature in the cryptocurrency market and is one of the highlights of this type of asset. Some factors that contribute to volatility in cryptocurrencies include:

1. **New Technology:** Cryptocurrencies are a relatively new technology, and the market is still developing and evolving. Uncertainty about their mass adoption and market maturity can lead to significant fluctuations in prices.

2. **Lack of Regulation:** The Cryptocurrencies is in

much of it deregulated compared to other financial markets. Lack of regulation can lead to speculation and price manipulation, contributing to volatility.

3. **Market Sentiment:** Investors' emotions and sentiment can influence the behavior of the cryptocurrency market. Positive or negative news, comments from public figures or unexpected events can cause sharp movements in prices.

4. **Limited Liquidity:** Some cryptocurrencies may have limited liquidity, meaning there are fewer market participants and less trading volume. This can lead to more extreme movements in prices.

5. **24/Market:** Unlike traditional financial markets, the cryptocurrency market operates 24/7. This allows events and news that occur at any time to quickly influence prices.

Volatility in cryptocurrencies can present both opportunities and risks for investors. On the one hand, volatility can provide the opportunity to make significant gains in a short period of time. On the other hand, it can also lead to substantial losses if prices fall rapidly.

It is important for investors to understand and manage the risks associated with volatility in cryptocurrencies. Diversify investments, set loss limits and maintain a long-term focus Term are Some of the strategies that can help mitigate the effects of volatility and make more informed decisions in the cryptocurrency market.

You Should Also Consider CBDCs

CBDCs (Central Bank Digital Currencies) are digital currencies issued and backed by countries' central banks. Unlike decentralized cryptocurrencies like Bitcoin, which are not controlled by any central authority, CBDCs are issued and regulated by each country's central financial institutions.

CBDCs are created as a digital evolution of traditional fiat currency, such as the US dollar, euro or Japanese yen. It aims to provide a secure and efficient digital way to conduct financial transactions and electronic payments, while maintaining stability and trust in the financial and monetary system.

There are two main models of CBDC:

1. **Wholesale Model:** This model focuses on providing financial institutions (commercial banks and other financial institutions) with access to a CBDC that is used for interbank settlements and high-value payments. This approach seeks to improve the efficiency and reduce the costs of settlement and clearing systems between banks.

2. **Retail Model:** In this model, CBDC Be

Available directly to the general public, which would allow individuals and businesses to make everyday payments and transactions using the central bank-issued digital currency. This would offer a digital alternative to traditional forms of money and facilitate the use of secure and efficient electronic payments.

Some potential benefits of implementing CBDC include:

- Greater financial inclusion: CBDCs could provide access to financial services to people who are unbanked or have limited access to traditional banking services.
- Greater efficiency and speed in payments: CBDC transactions could be faster and more efficient than traditional transfers and cash payments.
- Reduced risk of fraud and theft: CBDCs can incorporate advanced security technologies, which could reduce the risk of fraud and theft.

However, there are also important challenges and considerations for the implementation of CBDCs, such as privacy and security of personal data, protection against money laundering and terrorist financing, and interoperability between different CBDC systems in the international arena.

At present, several central banks around the world are exploring and experimenting with the issuance of CBDCs. Some countries, such as China, have made considerable progress in implementing their CBDCs, while others are conducting pilot tests and studies to assess the feasibility and potential benefits and risks of adopting this technology.

We must also talk about stable coins or stable currencies

Stablecoins are a class of cryptocurrencies designed to maintain a stable value and not experience the same volatility as many other cryptocurrencies, such as Bitcoin and Ethereum. Its main objective is to provide a digital alternative to traditional fiat currencies (such as the US dollar or euro) that can be used to transact in the world of cryptocurrencies and the digital economy.

There are three main types of stablecoins:

1. **Asset-backed stablecoins:** These stablecoins are backed by a reserve of physical assets, such as fiat currencies (dollars, euros, etc.) or precious metals (gold, silver). For every stablecoin issued, there is an equivalent amount of physical assets reserved in a bank account or safeguarded.

2. **Crypto collateralized stablecoins:** These Stablecoins use cryptocurrencies as collateral to back their value. Users lock an amount of cryptocurrencies (such as Ethereum) into a smart contract as collateral and in return, receive stablecoins. The amount of

cryptocurrency locked as collateral must exceed the Value of stablecoins issued to maintain stability.

3. **Algorithmic stablecoins:** These stablecoins are not backed by physical assets or cryptographic collateral, but instead use algorithms and supply and demand mechanisms to keep their value stable. If the price of the stablecoin falls below its target value, more coins are issued to increase the supply and reduce the price. If the price rises above the target value, coins are burned to reduce supply and increase the price.

Stablecoins have several important uses and advantages:

- **Price stability:** Having a value pegged to a fiat currency or physical asset, stablecoins offer greater price stability compared to other volatile cryptocurrencies. This makes them useful as a medium of exchange and storage of value in environments where volatility is undesirable.

- **Facilitate cryptocurrency transactions:** Stablecoins allow users to transact in cryptocurrencies without having to worry about price fluctuations that often affect other cryptocurrencies.

- **Facilitate access to DeFi:** Stablecoins are widely used in the decentralized finance (DeFi) ecosystem as a means to provide liquidity and participate in lending and other financial services without being exposed to the volatility of other cryptocurrencies.

It is important to note that although stablecoins strive to maintain a stable value, they are not without risks and challenges. Some asset-backed stablecoins may face auditing issues and insufficient reserves, while algorithmic stablecoins may face difficulties in maintaining their target value in conditions of extreme volatility. Users should research and understand the mechanisms behind each type of stablecoin before using them in their financial transactions and operations.

Air drops or free distribution if you read correctly; FREE

In the context of cryptocurrencies, an "airdrop" refers to the free distribution of tokens or cryptocurrencies to many people. It is a marketing strategy used by crypto projects to increase the visibility and adoption of their tokens, as well as to reward existing users or investors.

The airdrop process involves tokens being distributed to certain recipients at no cost to them. These recipients can be holders of a specific cryptocurrency, members of a community, or users who meet certain criteria. established by the project.

An airdrop can occur in different ways, such as:

1. Snapshot Airdrop: The project takes a snapshot of the blockchain of an existing cryptocurrency at a specific date and time. The tokens of the new project are then distributed proportionally to the holders of the cryptocurrency in the snapshot.

2. Participation in communities: Users who are part of online communities, such as Telegram groups, forums or social networks, can receive free tokens as a form of appreciation for their active participation.

3. Platform registration: Sometimes, projects can perform airdrops for those who register on their platform or subscribe to their newsletters.

4. Specific tasks or challenges: Projects can ask users to perform certain tasks or complete challenges, such as sharing posts, referencing other users, or filling out forms, in order to receive airdrop tokens.

It's important to note that not all airdrops are legitimate, and some may be scamming attempts or ways to collect personal data. Therefore, it is always advisable to research the project and verify the authenticity before participating in any airdrop. In addition, some countries may have specific regulations regarding the Free token distribution, so it's crucial to make sure you comply with local laws before participating in an Airdrop.

There is also something called Dark pool.

A "dark pool" in cryptocurrencies refers to a platform or market where investors can transact privately and without information about those operations being fully visible to the general public. Dark pools are a way of trading cryptocurrencies that offer greater privacy and discretion to participants.

In a dark pool, orders to buy and sell cryptocurrencies are not publicly displayed in an order book as in traditional markets. Instead, orders are executed internally within the dark pool without the details being visible to other participants or to the market.

These private trading platforms are used by institutional investors and professional traders looking to avoid the impact their large orders could have on market prices if executed on a public and visible exchange. The dark pool allows them to maintain the confidentiality of their trades and minimize price slippage.

However, the use of dark pools has also been the subject of controversy, with some critics arguing that they can increase the lack of transparency in the market and allow abuse or manipulation of prices. As a result, some financial regulators are evaluating the regulation of dark pools in the cryptocurrency world to ensure a fair and transparent market.

In short, a dark pool in cryptocurrencies is a private trading platform that allows investors to operate discreetly and confidentially, avoiding public exposure of their buy and sell orders.

11 WHERE TO SEE CRYPTO CHARTS

You can view cryptocurrency candlestick charts on various platforms and websites specializing in price analysis and cryptocurrency market data. Some of the most popular and reliable options are:

1.	CoinMarketCap: CoinMarketCap is one of the most well-known websites for cryptocurrency information, including candlestick candlestick price charts.

CoinMarketCap is one of the most popular and used websites in the world of cryptocurrencies. It was launched in 2013 and has become one of the leading sources of information for real-time data on cryptocurrencies, including prices, market capitalization, trading volume, circulating supply, news, and other relevant data.

Functions and features of CoinMarketCap:

1. Cryptocurrency Listing: CoinMarketCap displays an extensive list of cryptocurrencies, including the most popular and the least known, with detailed information about each of them.

2. Classification by market capitalization: Cryptocurrencies are classified according to their market capitalization, which is the total value of all coins in circulation multiplied by their price.

3. Charting and Analysis: CoinMarketCap provides price charts with candlestick candlesticks and other technical analysis tools to help users understand market behavior.

4. News and Information: The platform offers news and updates about the world of cryptocurrencies to keep users informed about important events and developments.

5. Tracking tools: Users can create custom lists of cryptocurrencies to follow their favorite assets and receive notifications of significant changes in the market.

Launch of new cryptocurrencies:

The launch of new cryptocurrencies can occur in various ways and depends on the project and its specific goal. Some of the common ways to launch new cryptocurrencies are:

1. Initial Coin Offering (ICO): It is a way to raise funds for a new crypto project by selling its tokens before it is officially launched. Investors buy the tokens in the hope that their value will increase once

cryptocurrency is in circulation.

2. Independent blockchain launch: Some cryptocurrencies, such as Bitcoin and Ethereum, were launched with their own blockchain and protocol from scratch, allowing them to have an independent network.

3. Forks: A fork is an update or modification in the source code of an existing cryptocurrency that gives rise to a new cryptocurrency. They can be hard forks or soft forks.

4. Airdrops: As explained above, some projects distribute new tokens or cryptocurrencies to certain recipients for free as a marketing strategy.

It is important to carefully research and evaluate any new cryptocurrency before investing in it, as the market is full of projects with varying goals and degrees of legitimacy. CoinMarketCap is a useful tool for gaining insights and conducting analysis before making investment decisions in the world of cryptocurrencies.

1. CoinGecko: CoinGecko is another platform that provides detailed data and analysis on cryptocurrencies, including candlestick candlestick charts.

CoinGecko is an online platform that

provides Detailed information and analysis of cryptocurrencies. Like CoinMarketCap, CoinGecko is widely used by the crypto community to obtain data.

accurate and up-to-date on the cryptocurrency market. Below are some of the key functions and features of CoinGecko:

1. Cryptocurrency Listing: CoinGecko offers an extensive list of cryptocurrencies, including both the most popular and the least known. The platform lists thousands of tokens and digital currencies with complete details about each of them.

2. Ranking and metrics: Cryptocurrencies are ranked based on various criteria such as market capitalization, trading volume, price, circulating supply, and other relevant metrics. CoinGecko uses a proprietary formula called "Gecko Ranking" to rank cryptocurrencies.

3. Charting and Analysis: CoinGecko provides interactive price charts with candlestick candlesticks and other technical analysis tools. Users can adjust time intervals and use various tools to perform advanced technical analysis.

4. Rating and Community: The platform allows users to rate and review cryptographic

projects. Users can also participate in online discussions and communities associated with each cryptocurrency.

5. Developer Information: CoinGecko offers details on the development teams behind each cryptocurrency, including their profiles and links to relevant websites and social media.

6. New cryptocurrencies and updates: CoinGecko provides information on new cryptocurrencies, forks, updates, and relevant events in the crypto space.

7. Additional Indices and Data: The platform offers additional indices and data on the cryptocurrency market, such as the "DeFi Index" and the "Bitcoin Dominance Index".

8. Investor Tools: CoinGecko provides useful tools for investors, including conversion calculators, custom portfolio trackers, and other utilities.

Overall, CoinGecko is a comprehensive and user-friendly platform that offers a wide range of information and analysis about the cryptocurrency market. Investors and cryptocurrency enthusiasts can use it as a valuable tool to make informed decisions and closely follow the dynamic world of cryptocurrencies.

3. TradingView: TradingView is a financial charting platform that offers a wide variety of tools for analyzing and visualizing cryptocurrency data.

1. Binance: If you have an account on cryptocurrency exchange Binance, you can also access charts
2. Binance is one of the largest and most popular cryptocurrency exchanges in the world. It offers a wide variety of services and features for users of different levels of experience. Below are some of the key functions and features of Binance:

1. Wide selection of cryptocurrencies: Binance offers an extensive list of cryptocurrencies for exchange, allowing users to trade a wide range of digital assets.

2. Competitive fees: Binance stands out for having low fees compared to other exchanges, making it an attractive option for those looking to reduce the costs of their operations.

3. Advanced Trading: Binance offers an advanced trading platform for experienced users who wish to perform technical analysis and use professional tools.

4. Intuitive Interface: In addition to the advanced trading platform, Binance also offers a simple and intuitive interface for beginner users, making the process of buying and selling cryptocurrencies easier.

5. Leverage Trading: Binance allows leverage trading for those looking to amplify their market positions and increase profit potential.

6. Futures contracts: Binance users can trade futures contracts for cryptocurrencies, allowing them to speculate on the price of long-term and short-term assets.

7. Staking and Loans: Binance offers staking services, allowing users to earn rewards for holding certain cryptocurrencies on their platform. It also allows users to lend their digital assets to earn interest.

8. Binance Launchpad: Binance Launchpad is a platform that allows crypto projects to raise funds through initial coin offerings (ICOs) and token exchange offerings (IEOs).

9. Binance Debit Cards: Binance offers debit cards that allow users to spend their cryptocurrencies directly at merchant establishments.

10. Security: Binance strives to maintain high

security standards and offers two-factor authentication (2FA) to protect user accounts.

11. Mobile Apps: Binance offers mobile apps for iOS and Android devices, allowing users to trade and access their accounts from anywhere.

These are just some of the functions and features that Binance offers. As with any financial platform, it is important to fully research and understand the services and associated risks before using it to trade or invest in cryptocurrencies.

5. Kraken: Kraken is another exchange that offers advanced candlestick charting for various cryptocurrencies.

These are just a few options, and there are plenty of other platforms and apps available for viewing cryptocurrency candlestick charts. Always remember to make sure you use reliable and recognized sources to get accurate and up-to-date data on the cryptocurrency market.

Crypto domains:

Crypto domains, also known as address aliases, are a convenient way to simplify and streamline the process of sending and receiving cryptocurrency.

Instead of using a long string of alphanumeric characters representing a wallet's address, crypto domains make it possible to assign a readable, easy-to-remember name to that address.

For example, instead of sending funds to the address "1JXYRBuwy6j4PFfycJpXTB2Q4tR1LQtVu", you can use the domain "mywallet" to make the transaction. This makes the process of sending and receiving cryptocurrencies more user-friendly and accessible to users, especially those who are not familiar with complex cryptocurrency addresses.

Crypto domains can be registered through specialized services that offer this functionality. Each cryptocurrency or blockchain may have different domain providers, and some projects even offer their own address alias solutions.

It is important to note that although crypto domains simplify the process of sending and receiving funds, security is still of paramount importance. It is crucial to use reliable services and providers and maintain the privacy of the wallet private key associated with the domain.

Regarding the possibility of sending different types of cryptocurrencies to the same wallet, this is possible thanks to the interoperable nature of many wallets and platforms. Modern wallets and exchanges often support multiple cryptocurrencies and tokens in a single direction.

This is achieved through the adoption of common standards, such as the ERC-20 protocol on Ethereum, which allows tokens created on the Ethereum network To be easily integrated into compatible wallets and exchanges. Also, some wallets and platforms Allow Add multiple accounts or addresses of different cryptocurrencies within the same interface, which facilitates the tracking and management of funds.

However, it is important to note that some cryptocurrencies and tokens may have specific requirements or unique characteristics, so it is essential to read and understand the conditions of each project before sending funds.

In short, crypto domains are a convenient tool to simplify the process of sending and receiving cryptocurrencies, providing a more accessible experience for users. In addition, the ability to send different types of cryptocurrencies to the same wallet is achieved thanks to interoperability and common standards adopted by many crypto wallets and platforms. However, security remains paramount at all times to protect digital assets and ensure an optimal experience in the world of cryptocurrencies.

And of course, here are some examples of companies that provide crypto domain services:

1. Unstoppable Domains: Unstoppable Domains is a company that offers crypto domain services on multiple blockchains, including Ethereum and Zilliqa. Allow users to register .crypto and .zil domains, which are used to simplify cryptocurrency and token addresses.

2. MyEtherWallet (MEW): MEW is a popular Ethereum wallet that also offers crypto domain services through its ENS (Ethereum Name Service) program. With ENS, users can assign readable names to Ethereum addresses and receive payments through these names.

3. Handshake: Handshake is a blockchain project that seeks to decentralize and democratize the domain name infrastructure on the Internet. It allows users to register domains on the Handshake blockchain and use them for web services and cryptocurrency addresses.

4. Ethereum Name Service (ENS): ENS is a system of crypto domains on the Ethereum network. It allows users to register .eth domains and associate them with their Ethereum addresses, making it easier to send and receive funds.

5.   Zilliqa Name Service (ZNS): ZNS is a crypto domain system on the Zilliqa network, which allows users to register .zil domains on the Zilliqa blockchain and associate them with their Zilliqa addresses.

1.   Uniregistry: Uniregistry is a platform that allows users to register crypto domains, including popular extensions such as .crypto, .eth, and .zil. They also offer brand protection and domain management services.

7. Namebase: Namebase is a crypto domain platform on the Handshake blockchain. It allows users to register .hns domains and use them for cryptocurrency addresses and web services.

These are just a few of the companies that offer crypto domain services. Each has its own characteristics and approaches, but they all have in common the goal of simplifying and streamlining transactions and services in the crypto ecosystem. It is important to research and compare the options available to choose the one that best suits your needs and preferences.

12 BITCOIN ETF

Yes, an ETF (Exchange-Traded Fund) based on Bitcoin is possible. A Bitcoin ETF is an investment fund that is listed on a stock exchange and seeks to replicate the price performance of Bitcoin or some Bitcoin-related index.

Over the years, there have been several attempts by different companies and entities to launch a Bitcoin ETF in regulated financial markets, such as the United States. However, until the cut-off date of my knowledge in September 2021, the Commission of The U.S. Securities and Exchange (SEC) has rejected multiple Bitcoin ETF applications, citing concerns related to investor protection and market manipulation.

The main obstacle to the approval of a Bitcoin ETF has been the volatility and unregulated nature of the cryptocurrency market. The SEC has expressed concern about the lack of oversight and control in the cryptocurrency market, as well as the possibility of price manipulation and other associated risks.

Despite previous rejections, it is important to note that interest in a Bitcoin ETF remains high, and it is possible that in the future the SEC will approve an ETF if certain regulatory requirements are met and steps are taken to address existing concerns.

Meanwhile, there are other financial products in some international markets that allow investors to gain exposure to Bitcoin through instruments such as Bitcoin futures contracts and structured notes tied to Bitcoin performance. However, these products may not be available in all countries and may carry other associated risks and costs. Investors interested in gaining exposure to Bitcoin through financial instruments should research and fully understand the risks and benefits first. to make investment decisions.

Once Bitcoin ETFs (Exchange-Traded Funds) are approved and available to investors, they will be a great way to invest for those looking for a simpler and less complicated option to gain exposure to Bitcoin. Especially for older investors who are retiring or already retired and who may prefer to avoid buying and selling cryptocurrencies outright, Bitcoin ETFs offer several advantages:

1. Simplifying the investment process: Bitcoin ETFs

allow investors to access exposure to Bitcoin through their traditional broker-dealer, simplifying the investment process. It is not necessary to open accounts on cryptocurrency exchanges or worry about the custody and security of cryptocurrencies.

2. Professional management: With a Bitcoin ETF, investors can delegate the management of their Bitcoin investment to professional and experienced managers who manage the fund. This frees investors from the need to constantly monitor the market and make investment decisions.

3. Diversification: Bitcoin ETFs typically have a diversified basket of underlying Bitcoin-related assets, which can reduce the risk associated with the volatility of a single asset. This can be especially beneficial for older investors looking to protect their capital and generate stable income in retirement.

4. Lower risk of loss: By delegating management to professionals and having a diversified portfolio, investors can mitigate the risk of loss in comparison with investing in a single cryptocurrency.

5. Liquidity and ease of trading: Bitcoin ETFs are traded on regulated exchanges, providing liquidity and ease of trading. Investors can buy and sell shares in the ETF at market prices at any time during trading hours.

It is important to note that although Bitcoin ETFs can offer several advantages for older investors, they also carry inherent risks associated with the volatility of the cryptocurrency market and fund management. As with any investment, it is essential that investors are properly informed, understand the risks, and consult with financial advisors before making investment decisions. Once approved, Bitcoin ETFs can be an attractive option for those who wish to gain exposure to Bitcoin more easily and without the need to deal directly with buying and selling cryptocurrencies.

On the other hand, there are also bitcoin futures contracts.

Bitcoin-based futures contracts are a type of financial instrument that allows investors to speculate on the future price of Bitcoin. In essence, a Bitcoin futures contract is an agreement between two parties to buy or sell Bitcoin at an agreed price at a certain future date. These contracts are traded on specialized markets, such as cryptocurrency futures exchanges.

Here are some key points to better understand Bitcoin-based futures contracts:

1. Operation: A Bitcoin futures contract specifies the price at which it is agreed to buy or sell a specific amount of Bitcoin on a given future date. Futures contracts can be settled in cash or in physical Bitcoin,

depending on the terms of the contract.

2. Price Speculation: Investors can use Bitcoin futures contracts to speculate on whether the price of Bitcoin will rise or fall in the future. If they believe that the price will increase, they can buy futures contracts (long position); If they believe the price will decrease, they can sell futures contracts (short position).

3. Leverage: Bitcoin futures contracts often allow leverage, meaning that investors can control a larger amount of Bitcoin with a relatively small investment. However, leverage also carries a higher risk of losses.

4. Daily settlement: In many futures markets, contracts are settled daily to reflect the day's gains and losses. This means that investors must maintain sufficient capital. in your account to cover daily price fluctuations.

5. Hedging: Bitcoin futures contracts can also be used by companies and institutional investors to hedge their risks associated with Bitcoin price volatility. This allows them to protect themselves against movements unfavorable in the market.

6. Expiration Date: Futures contracts have a specific expiration date on which they must be settled. Some contracts may have monthly, quarterly, or even annual maturities.

Importantly, Bitcoin futures contracts are complex financial instruments and carry a high degree of risk. The volatility of the Bitcoin market can lead to significant gains, but it can also lead to substantial losses. Investors interested in trading Bitcoin futures should have a solid understanding of how they work and be willing to take the associated risks. It is advisable to seek financial advice before trading Bitcoin futures contracts.

13 DO NOT SEND YOUR CRYPTO TO ANOTHER UNIVERSE

That's how you can lose your crypto if you send it wrong and you will not be able to recover it

Here are some examples of how you can lose it without wanting or having planned it and you can only be left with a bad memory and clear the experience of having sent it to nobody knows where

Losing cryptocurrencies due to errors in the receiving address is a common and potentially costly problem in the cryptocurrency world. Here are some examples of how this can occur:

1. Typographical error in the receiving address: If when sending cryptocurrency, you make a simple typo in the receiving address, your funds may end up in the wrong address and be unrecoverable. For example, if the correct address is "1AbCdEfGh..." and you accidentally type "1aBcDeFgH...", the funds will be sent to an unknown wallet and you will not be able to recover them.

2. Send to an unsupported address: Some cryptocurrencies have address formats different, and some wallets are not compatible with certain currencies or tokens. If you try to send a cryptocurrency to an unsupported address, the funds may be lost in cyberspace with no possibility of recovery.

3. Send to an inactive or lost wallet: If you send cryptocurrency to a wallet that is inactive or lost, the funds will be stored at that address without being able to access them.

4. Send to an incorrect exchange address: If you send Cryptocurrencies to an incorrect or outdated exchange address, the funds may be blocked on the exchange platform, and you may need to contact customer support to try to recover them.

5. Phishing or scams: In some cases, scammers may create fake websites or emails that resemble

legitimate platforms to steal your funds. If you provide your receiving address to a malicious site, you could lose your cryptocurrency.

To avoid these situations and minimize the risk of losing your cryptocurrencies, it is essential to follow some security practices:

- Always verify the receiving address twice before making a transaction.
- Use copy and paste to enter addresses, instead of typing them manually.
- Check that the wallet or exchange is compatible with the cryptocurrency you want to send.
- Keep your devices and wallets up-to-date and protected with security measures like two-factor authentication (2FA).
- Be wary of suspicious links and emails and be sure to access crypto platforms through official URLs.

Remember that cryptocurrency transactions are irreversible, so it is essential to be cautious and cautious when sending funds. It's always best to research and educate yourself on safe practices to protect your digital assets.

You can also have a problem if you lose your cell phone and there is access to your cryptos from there and you do not have a good protection key.

Losing your cryptocurrencies through your cell

phone can be an unfortunate and costly situation. Here are some of the most common dangers associated with losing cryptocurrency via cell phone:

1. Theft or loss of the phone: If your cell phone is lost or stolen, someone else could have access to your crypto wallet if it is not properly protected. Without proper security measures, such as passwords or two-factor authentication (2FA), a third party could access your wallet and transfer your cryptocurrencies to another address, which would result in a permanent loss of your funds.

2. Malware and phishing: Cell phones are susceptible to malware and phishing attacks. Hackers can use malicious apps or fake websites to steal your credentials and gain access to your crypto wallet. They may also send fraudulent emails or text messages that lead you to reveal sensitive information.

3. Improper backups: If you have a backup of your crypto wallet on your cell phone, it is important to make sure it is protected and encrypted correctly. If someone else has access to your phone or unprotected backups, they could gain access to your private keys and control your cryptocurrency.

4. Technical glitches or device damage: Technical glitches on your cell phone or physical damage to the device could result in loss of access to your crypto wallet and therefore your cryptocurrencies. If you

don't have a secure, up-to-date backup, you could permanently lose access to your funds.

5. Unsecured third-party apps: By using third-party apps to manage your cryptocurrencies on your cell phone, you risk exposing yourself to security vulnerabilities. Some malicious apps can access your personal data or try to steal your cryptocurrency.

To protect your cryptocurrencies and avoid these dangers, here are some security measures you can take:

- Use reliable and well-reputed crypto wallets to store your funds.
- Enable additional security measures, such as two-factor authentication (2FA) and PIN to access your wallet.
- Back up your wallets on a regular basis and keep them stored in a safe place, such as an external storage device or paper wallet.
- Beware of suspicious links and emails and avoid clicking on unknown links or providing personal information to untrusted sources.
- Keep your cell phone updated with the latest software updates and use reliable security solutions to protect it against malware and attacks.

By taking precautions and following good security practices, you can significantly reduce the risk of losing your cryptocurrencies through your cell phone.

Remember that the security of your digital assets is a personal responsibility, and it is always best to be well informed and prepared to protect your cryptocurrency investments.

It is very important that you use only reliable applications to buy your crypto and as soon as possible download your crypto to your wallet cold your keys your crypto there nobody can touch them

And another very important thing keeps your phrases Of safe recovery because if you lose them you will most likely lose your crypto when you need your Fraces to execute a recovery

Recovery key phrases, also known as mnemonic phrases, seed words or seed phrases, are sequences of words that act as a backup form to regain access to your crypto wallet. These phrases are essential to protect your cryptocurrencies and make sure you can get your funds back in case you lose access to your wallet or device.

When you create a new crypto wallet, you are usually provided with a recovery key phrase of 12, 18, or 24 words. These words are generated randomly and follow a standard known as BIP39 (Bitcoin Improvement Proposal 39). Each keyword in the key phrase represents a specific, standard code, ensuring that you can use it to retrieve your cryptocurrencies in different BIP39-compatible wallets.

It is important to note the following about key recovery phrases:

1. **Importance of security:** The recovery key phrase is a critical piece of information and should be kept secret. Don't share it with anyone and never enter it on an untrusted website or platform.

2. **Secure backup:** You must write the recovery key phrase on paper or water and fire resistant material. Keep it in a safe place away from unauthorized access. Some people also use external storage devices or specialized metal wallets to store it.

3. **Accuracy and order:** The sequence of words must be entered in the correct order and without typographical errors. A single mistake in a word could prevent you from getting your funds back.

4. **Wallet Recovery:** If you lose your crypto device or wallet, you can restore your cryptocurrencies to a new wallet using the recovery key phrase. Most crypto wallets will ask you to enter the words in the correct order to restore your funds.

5. **One Chance:** Once you've used the recovery key phrase to restore your funds to a new wallet, it's important to generate a new passphrase and keep it safe.

Key recovery phrases are an essential part of security in the world of cryptocurrencies. By protecting and keeping this information secure, you ensure that you can recover your funds in case of any problems or loss of access to your wallet. Remember that security is your responsibility, so always take the right precautions to protect your digital assets.

Now in case you lose your cold wallet this is what you should do (as long as you have the recovery keys) it is not advisable to save them in the same place.

If you lose your cold wallet or hardware wallet, you can recover your cryptocurrencies using the recovery keys, also known as phrase mnemonic or seed phrase. Here's a step-by-step guide to doing so:

1. **Stay calm and confirm the loss**: The first thing is to stay calm and confirm that you have really lost your cold wallet or hardware wallet. Make sure you search in every possible place and make sure you can't access your device.

2. **Get a new hardware wallet**: If you don't already have another hardware wallet, purchase a new one from a trusted source. It is essential to make sure that it is new and has never been used by someone else.

3. **Set up the new hardware wallet**: Follow the manufacturer's instructions to set up your new

hardware wallet. During setup, you'll be asked to enter your recovery phrase.

4. **Enter the recovery phrase**: In this step, you need to enter the recovery phrase that you saved when you set up your original cold wallet. Words should be entered in the correct order and without typos.

5. **Restore your cryptocurrency**: Once you enter the recovery phrase, the new hardware wallet will recover all accounts and funds associated with that phrase. The cold wallet will now contain your cryptocurrencies and you will be able to access them as you did with the previous wallet.

It is important to keep in mind some aspects to maintain security during this process:

- Never share your recovery phrase with anyone else. It is your private key to access your cryptocurrencies.
- Make sure no one else is watching while you enter your recovery phrase.
- Keep the recovery phrase in a safe and secure place. Do not store it online or on electronic devices vulnerable to hacking or loss.

Remember that the recovery phrase is the key to accessing your funds, so try to always keep it secret and secure. With the hardware wallet and recovery phrase properly protected, your cryptocurrencies will be safe even if you lose or damage your original cold wallet.

14. WHERE DO I BUY CRYPTO?

There are numerous companies where you can buy cryptocurrencies, and each has its own advantages and disadvantages. Here are some examples of the best companies both centralized and decentralized:

Centralized Companies:

1. Coinbase: It is one of the largest and most popular cryptocurrency platforms in the world. It offers a beginner-friendly interface and a wide variety of cryptocurrencies to buy and sell. Coinbase is known for its high security and is a popular choice for those just starting out in the world of cryptocurrencies.

2. Binance: It is one of the world's leading cryptocurrency exchanges in terms of trading volume. It offers a wide variety of cryptocurrencies and advanced trading features. Binance also boasts its own cryptocurrency, Binance Coin (BNB), which is used to get discounts on transaction fees.

3. Kraken: It is a cryptocurrency platform with a strong reputation for security and regulatory compliance. It offers a wide variety of cryptocurrencies and advanced trading features. Kraken is also known for its excellent customer service.

1. Bitstamp: It is one of the oldest exchanges in the world and has maintained a solid reputation over the years. It offers a selection of popular cryptocurrencies and is known for its focus on security.

4 Uphold is a financial company and cryptocurrency platform that was founded in 2014

with the aim of facilitating access to a wide variety of digital and traditional assets. Uphold's platform allows users to buy, sell, store and send a wide range of cryptocurrencies, as well as fiat currencies and precious metals.

Features and Functions of Uphold:

1. Asset Diversity: Uphold offers a wide variety of popular cryptocurrencies, including Bitcoin, Ethereum, Litecoin, Ripple (XRP), and many more. It also supports fiat currencies such as the U.S. dollar, euro, British pound, and others, as well as precious metals such as gold, silver, and platinum.

2. Virtual and Physical Cards: Uphold offers virtual and physical debit cards that allow users to spend their cryptocurrencies and fiat currencies anywhere that accepts debit cards.

3. Easy Conversion: Users can make instant conversions between different cryptocurrencies and fiat currencies within the platform, making it easy to manage and exchange assets.

4. Transparent Commissions: Uphold is known for its focus on transparency and clarity of fees. Users can view the fees applied before making any transaction.

5. Security: Uphold uses advanced security measures, such as two-factor authentication (2FA) and

data encryption, to protect users' accounts and assets.

6. Regulations and Compliance: Uphold operates under strict regulatory standards and complies with applicable financial laws in the jurisdictions where it operates.

7. Bank Card Integration: Uphold users can link their bank accounts and credit/debit cards to deposit and withdraw funds from the platform.

Importantly, Uphold has sought to offer an accessible and simple service for users from different parts of the world, allowing them to access a wide range of financial assets. Its focus on diversification and ease of use has attracted a large number of users, from beginners to more experienced investors.

However, as with any financial platform, it is crucial that users research and fully understand the fees, policies, and features before using Uphold or any other cryptocurrency platform. Research and prudence are critical to making informed decisions in the world of cryptocurrencies and digital finance.

Decentralized Companies:

1. Uniswap: It is a decentralized exchange (DEX) based on the Ethereum network. It allows users to exchange cryptocurrencies directly and without intermediaries. Uniswap uses smart contracts and

allows users to stake their cryptocurrencies for returns.

2. SushiSwap: It is another Ethereum-based DEX that allows users to trade cryptocurrencies in a decentralized way. SushiSwap offers a variety de funciones de yield farming y staking.

3. PancakeSwap: It is a DEX based on the Binance Smart Chain (BSC) network. Offers a similar experience to Uniswap but with lower transaction fees due to the BSC chain.

4. 1inch: It is a DEX aggregator that allows users to get the best rates from multiple decentralized exchanges. 1inch uses algorithms to find the most efficient and economical exchange routes for users.

The key difference between centralized and decentralized companies lies in the ownership and control of assets. In centralized companies, users must deposit their funds on the platform and trust the security and custody of the company. This may be easier to use for beginners, but it involves a security risk as the funds are kept in the hands of third parties.

Instead, in decentralized companies, users maintain full control of their assets in their personal wallets and transactions are made directly through smart contracts. This gives users greater privacy and security as they do not need to trust third parties with their funds. However, using decentralized platforms may require a bit more technical knowledge and may have higher transaction fees due to the nature of the blockchain.

In short, both centralized and decentralized companies have their advantages and disadvantages. The choice will depend on the preferences and individual needs of each user. It is important to research and understand how these platforms work before making investment and trading decisions in the world of cryptocurrencies.

"Dollar cost averaging" (DCA), also known as dollar cost averaging, is an investment strategy in cryptocurrencies (and other financial assets) that involves making regular purchases of a certain amount of cryptocurrencies at fixed time intervals, regardless of the current price of the asset. Instead of trying to predict market movements and making large investments in a single moment, DCA relies on buying steady amounts of cryptocurrencies at regular intervals, regardless of whether the price goes up or down.

The DCA process can be explained with the following steps:

1. Set a plan: The investor decides the amount of money he is willing to invest in cryptocurrencies and the time interval in which he will make regular purchases. For example, you might decide to invest $100 each month in cryptocurrencies.

2. Buy at regular intervals: Regardless of the price of the Cryptocurrency in the market, the investor buys the fixed number of cryptocurrencies specified in his plan at regular intervals. For example, every month you buy $100 in cryptocurrencies, regardless of whether the price has risen or fallen since the previous purchase.

3. Benefits of DCA: DCA is a strategy that seeks to reduce exposure to market risk and smooth volatility. When the price of cryptocurrency is low, the investor acquires more units with his investment. And when the price is high, you buy fewer units. Ultimately, this strategy averages the cost of purchase at different points in the market and can help mitigate the risk of making a large investment at a time of high volatility.

It is important to note that DCA does not guarantee profits and does not eliminate the risk associated with investing in cryptocurrencies. However, it is a popular strategy among investors who want to maintain constant exposure to the cryptocurrency market and avoid trying to predict short-term market movements.

It should be noted that DCA is a long-term strategy and may be suitable for investors who are looking to maintain a long-term investment perspective and are not comfortable with the short-term volatility of the cryptocurrency market. As always, before adopting any investment strategy, it is important to research and understand the risks associated with cryptocurrencies and consider your own financial goals and risk tolerance.

15 BEWARE OF SCAMS

The risks of scams in the world of cryptocurrencies are a major concern for investors and users. Scammers use various deceptive tactics to steal funds and cryptocurrencies from unsuspecting people. Here are some of the most common risks and how you can avoid falling into them:

1. Phishing and Fake Websites: Scammers can create fake websites that resemble legitimate cryptocurrency exchanges or wallets. These sites trick users into entering their credentials and private keys, allowing scammers to access their funds. To avoid this, always make sure to access official websites and use secure connections (https://) before entering your details.

2. Investment Scams: Scammers may promise extraordinary returns on cryptocurrency investment through Ponzi schemes or fraudulent investment programs. It's important Remember that no investment guarantees constant and high returns. Research well before investing in any project and be wary of promises of quick and easy profits.

3. Fake Airdrop Offers or Gifts: Scammers may offer free cryptocurrency or airdrops in exchange for providing certain personal information or private keys. Never reveal your private keys to anyone and carefully verify the authenticity of any gift offer or airdrop.

4. Pump and Dump schemes: Scammers can create groups to manipulate the price of a little-known cryptocurrency and cause its price to rise. Then, they sell their coins at a higher price, leading to a drop in price (dump). Uninformed investors can suffer significant losses in these types of schemes.

5. ICO (Initial Coin Offering) Scams: ICOs are a way

to raise funds for new cryptocurrency projects. Some projects may be fraudulent or fail to deliver on their promises once they have raised funds. Thoroughly researching the project, the team behind it, and the viability of the product or service is crucial before investing in an ICO.

6. Wallet and App Scams: Some apps and mobile wallets can be malicious and designed to steal your private keys or access your funds. Be sure to download apps and wallets from trusted sources and check reviews and ratings before using them.

To protect yourself from scams and scams, it is essential to follow some key practices:

- Research and educate: Learn about cryptocurrencies and common scam tactics. The more informed you are, the less likely you are to fall into traps.

- Verify authenticity: Before providing any personal information or private keys, make sure the platform or service is legitimate and trustworthy.

- Use secure wallets: Use hardware wallets or well-reputed mobile wallets that offer high levels of security.

- Keep your private keys private: Never reveal your private keys to anyone and avoid sharing sensitive details online.

- Beware of overly good offers: Be wary of promises of quick and high profits, as they are usually signs of fraudulent schemes.

In short, the cryptocurrency industry offers exciting opportunities, but it is also exposed to scams and risks. The best defense is education and caution. Always do your research before investing in or providing personal information, and keep your private keys and funds in a safe place. With a good understanding of the risks and security measures, you can more safely and responsibly enjoy the world of cryptocurrencies.

16   ISO2022

ISO 2022 is an international standard developed by the International Organization for Standardization (ISO) that refers to a set of standards that define the character encoding for the representation of text in computer systems. It was first published in 1986 and has been revised and updated several times since.

The ISO 2022 standard aims to facilitate interoperability of computer systems in different countries and regions with different character sets and coding systems. This is especially important in the context of communication and data exchange between computer systems in different countries, which may use different alphabets and character sets to represent text.

ISO 2022 uses character set change techniques to enable the encoding of characters specific to different languages and writing systems. This is accomplished by including specific control sequences that indicate the character set that will be used to represent the following text.

For example, if a computer system needs to send a message that contains English text, followed by Japanese text, it can use ISO 2022. to switch from a character set that represents English (for example, ASCII) to a character set that represents Japanese (for example, JIS X 0208). This allows the receiver of the message to Correctly interpret the text in each language.

ISO 2022 is especially useful in multilingual environments and in applications that require the exchange of data in different languages. It is also relevant in systems that need to handle special characters and accents of different languages.

It is important to mention that while ISO 2022 has been widely used in the past, its adoption has declined in favor of other, more modern character encoding standards, such as Unicode. Unicode is a broader and more comprehensive standard that allows almost all characters from all languages and scripts to be represented in a single, universal character set. As the globalization of communications and computing has grown, Unicode has become the dominant standard for character encoding in modern computer systems.

The ISO 2022 standard does not define or specify specific cryptocurrencies that are accepted or used within its scope. ISO 2022 is a standard that focuses on character encoding and aims to facilitate the interoperability of computer systems by allowing the representation of text in different languages and character sets.

Cryptocurrencies are digital assets based on Blockchain technology and are mainly used as means of exchange and storage of value. Its adoption and Acceptance in computer systems are not regulated by the ISO 2022 standard but depend on agreements and business decisions between the parties involved.

However, some popular cryptocurrencies, such as Bitcoin (BTC) and Ethereum (ETH), have been used in different applications and systems that can follow the ISO 2022 standard for character encoding. For example, in applications that require multilingual text representation or that interact with international payment systems, cryptocurrencies can be used as a medium of exchange.

Importantly, the adoption of cryptocurrencies in different computer applications and systems depends on various factors, such as local regulation, stakeholder acceptance, and available technical infrastructure. As blockchain technology and cryptocurrencies continue to evolve, additional solutions and standards may be developed to facilitate their integration into computer systems more broadly.

However, there are already some Cryptocurrencies accepted and recognized in ISO2022 at present (July 2023) these are the following: QUANT, XRP, STELLAR, HEDERA, IOTA, XDC AND ALGORAND

The coins you mention (Quant, XRP, Stellar, Hedera, IOTA, XDC, and Algorand) share some similarities, but they also have significant differences. Below, I will explain some of the most prominent similarities between these cryptocurrencies:

1. Blockchain or Distributed Ledger Technology (DLT): Most of these cryptocurrencies, such as XRP, Stellar, Hedera, IOTA, XDC, and Algorand, use blockchain or distributed ledger technology (DLT) to

record and validate transactions. However, it should be noted that Quant is not a cryptocurrency itself, but a platform that allows interoperability between different blockchains and systems.

2. Focus on Payments and Transactions: Both XRP, Stellar, Hedera and XDC are designed with a focus on payments and value transfers. XRP and Stellar are used in solutions for cross-border payments and remittances, while Hedera and XDC have similar applications in the realm of payments and transfers.

3. Scalability and Efficiency: Hedera and Algorand are known for their approaches to scalability and efficiency. Hedera uses a Proof of Stake (PoS)-based consensus algorithm and hashgraph technology to achieve high throughput and low latency. Algorand uses a consensus algorithm based on Pure Proof of Stake (PPoS) to achieve fast and decentralized transaction processing.

4. Internet of Things (IoT): IOTA is a cryptocurrency which has stood out for its focus on the Internet of Things (IoT). It uses a technology called Tangle instead of blockchain, allowing it to be scalable and efficient for exchanging data and value across devices.

It is important to mention that although these cryptocurrencies have some similarities, they also have their own specific characteristics and use cases. Each of them is developed to address different issues and applications within the crypto ecosystem. Before investing or using any of these cryptocurrencies, it is essential to research and understand their characteristics and uses, as well as evaluate the potential risks and rewards.

Las DAPPS

DApps, or decentralized applications, are computer applications that work on a blockchain network or distributed ledger technology (DLT). Unlike traditional applications, DApps operate in a decentralized environment and use blockchain technology to ensure the security, transparency, and reliability of their operations.

The main features of DApps are:

1. Decentralization: DApps do not depend on a Centralized server to work. Instead, they use a distributed network of nodes on the blockchain to process and validate transactions.

2. Open source: The source code of DApps is generally publicly available, which means that anyone can access and examine the code to verify its security and functionality.

3. Consensus: DApps use consensus mechanisms,

such as Proof of Work or Proof of Stake, to validate and confirm transactions on the network.

4. Transparency: All transactions and operations in a DApp are recorded on the blockchain and are visible to all network participants.

5. Security: The decentralized architecture and the use of advanced cryptography in DApps make them more resistant to attacks and manipulations.

6. Interoperability: Some DApps are designed to interact and collaborate with other applications and protocols on the same blockchain network.

DApps have a wide range of use cases, ranging from financial applications and smart contracts to gaming, e-commerce platforms, digital identity systems, and much more. Some popular examples of DApps are Uniswap, Compound, CryptoKitties, and Decentraland.

It is important to note that, although the DApps son Decentralized in their design, their adoption and use can still be linked to centralized platforms and services, such as cryptocurrency wallets, block explorers and blockchain network access applications. Full decentralization is a desirable goal for many DApps, but in today's reality, there may be some interaction with centralized components.

Joel Ochoa

Conclusion

In "Cryptocurrencies: The Digital Monetary Revolution"

We have made an exciting journey through the fascinating world of cryptocurrencies. From the beginnings of Bitcoin to exciting new innovations in the crypto ecosystem, we've explored how these technologies are transforming the way we perceive and use money.

We began our journey with the creation of Bitcoin and its powerful whitepaper, which presented the world with a decentralized digital monetary system without intermediaries. We discover how blockchain technology is the fundamental pillar that ensures the transfer of value in a transparent and immutable way. The concepts of cryptography and mining became our allies in understanding the functioning of cryptocurrencies.

As we go through the book, we delve into the vast universe of cryptocurrencies. We explore the various altcoins and tokens that have emerged over time, each with its own characteristics, use cases, and applications. We understood the importance of diversification in a crypto portfolio, enabling investors to adapt to an ever-changing market and discover new growth opportunities.

One of the most exciting aspects we explored was the rise of NFTs, Non-Fungible Tokens. We discover how they have revolutionized the way we value and market digital art, collectibles and unique assets, opening up a whole new world of possibilities for artists and collectors.

Continuing our journey, we dive into the impact of cryptocurrencies on the global economy. From their integration into the financial sector to their role in international trade, we understood how cryptocurrencies are challenging traditional norms and offering innovative solutions for an increasingly digital economy.

We cannot fail to mention the importance of security in the world of cryptocurrencies. Throughout the book, we highlight the need to protect our private keys and recovery phrases to prevent the loss of our digital assets. We also address the risks and precautions we should take when investing and trading in the cryptocurrency market.

In addition, we explore the different options for storing and protecting our cryptocurrencies, from cold wallets and hardware wallets to mobile wallets and secure exchanges. We understood the importance of education and research before Make financial decisions in this ever-evolving world.

As we go through the book, we also explore crypto credit and debit cards, which are facilitating the mass adoption of cryptocurrencies in everyday life. These cards allow users to spend their cryptocurrencies in physical establishments and online, providing a smoother and more accessible experience for cryptocurrency enthusiasts.

In addition, we discuss the importance of regulation and the role that governments and financial institutions play in the adoption and acceptance of cryptocurrencies in society. The relationship between crypto and regulatory authorities has been a relevant topic on this journey, and we understood the importance of a balanced approach to foster innovation and protect users.

In the financial realm, we explore how cryptocurrencies offer unique opportunities to invest and diversify our portfolios, providing access to global markets and new asset classes. However, we also recognized the risks associated with volatility and speculation, underscoring the importance of a sound investment strategy and a deep understanding of the projects we invest in.

On our journey, we also come across concepts such as stacking, farming and DeFi (Decentralized Finance), which have opened up new ones. horizons for users and provided opportunities to earn returns and actively participate in the economy Cryptographic.

Finally, we discuss the role cryptocurrency exchanges and platforms play in the ecosystem. From

CoinMarketCap and CoinGecko, which provide vital pricing information and market data, to popular exchanges such as Binance and Coinbase, which offer cryptocurrency trading and buying and selling services, we understood the importance of these platforms in accessing and participating in the world of cryptocurrencies.

In conclusion, "Cryptocurrencies: The Digital Monetary Revolution" has been an exciting journey through the innovative and disruptive technology that is changing the way we perceive and use money. From their creation to their mass adoption, cryptocurrencies have demonstrated their potential to transform the global economy, empower individuals, and provide opportunities for greater financial inclusion.

However, we also recognize that this journey has only just begun. The world of cryptocurrencies continues to evolve and face challenges, from regulation to security and scalability. But with a solid foundation of knowledge, education and a deep understanding of the risks and opportunities, enthusiasts and investors can responsibly participate in this exciting digital monetary revolution.

As we move into the future, we will continue to explore and learn about cryptocurrencies and their applications in our lives. With an open mind, due diligence and a responsible attitude, we can make the most of the opportunities this exciting world has to offer and be part of a new era in financial and technological history.

Let this book be a starting point for an exciting journey into the future of cryptocurrencies and the digital monetary revolution!

ABOUT THE AUTHOR

Joel Ochoa   one more enthusiast in this extraordinary world of cryptocurrencies

www.ingramcontent.com/pod-product-compliance
Lightning Source LLC
Chambersburg PA
CBHW062320290526
45794CB00005B/1840